Beachcomber's Guide

FROM CAPE COD TO CAPE HATTERAS

HENRY KEATTS

Gulf Publishing Company
Houston, Texas

> To my mother and brothers Jim and Chuck,
> for support, encouragement, and love.

Copyright © 1995 by Gulf Publishing Company. All rights reserved. This book, or parts thereof, may not be reproduced in any form without permission of the publisher.

Gulf Publishing Company
Book Division
P.O. Box 2608 ☐ Houston, Texas 77252-2608

10 9 8 7 6 5 4 3 2 1

Printed in the United States of America

Library of Congress Cataloging-in-Publication Data

Keatts, Henry.
  Beachcomber's guide : from Cape Cod to Cape Hatteras / Hank Keatts.
    p. cm.
  Includes bibliographical reference (p.   ) and index.
  ISBN 0-88415-130-1
  1. Seashore biology—Atlantic Coast (U.S.)   2. Beach-combing—Atlantic Coast (U.S.)   I. Title.
  QH104.5.A84K435   1995
  574.974—dc20   95-12534
                                                                     CIP

DISCLAIMER

Information presented by the author on the edibility of marine organisms and the treatment of stings often has been obtained from reference material. The reader should show extreme caution in regard to collecting marine organisms for consumption and treating their stings or bites. Be sure identification of the species is without doubt and that the specimen is from an unpolluted area if collected for consumption. Some species of marine organisms are potentially toxic.

Contents

Acknowledgments

Without the contributions and cooperation of the following individuals and organizations, this book would not have been possible: Prof. John Black, Prof. David Cox, Prof. Kenneth Ettlinger, Aaron Hirsh, Prof. George Lomaga, Prof. Charles McCarthy.

For use of their photographs, I am indebted to Michael Casalino, Chip Cooper, John Cossick, Jack Darrell, Michael deCamp, Norman Despres, Luther C. Goldman, Mike Haramis, George Harrison, Dianne Huppman, James P. Mattisson, Jim Matulis, T. C. Maurer, Dave and Sue Millhouser, Theano Nikitas, and Bradley Sheard.

I thank Dr. Charles Haddad for taking me aloft to obtain aerial photos of Long Island's barrier beaches; the National Park Service; and the United States Department of the Interior, Fish & Wildlife Service.

As always, special thanks are due to George Farr for his editorial assistance.

Introduction

There is a pleasure in the pathless woods,
There is a rapture on the lonely shore,
There is society, where none intrudes,
By the deep sea, and music in its roar:
I love not man the less, but Nature more.

Lord Byron

This book is a practical guide to the major groups of flora (plants) and fauna (animals) likely to be encountered while beachcombing from Cape Cod to Cape Hatteras. It does not attempt to describe all species of marine organisms found in the study area, only the more common ones. This guide contains but a modest representation of the marine organisms to be found in the study area, but it should be sufficient to introduce a beachcomber to the major groups of inhabitants of the fascinating environment. Many of the plants and animals introduced in one habitat (rock jetty or sandy beach) exist in other habitats as well, but are usually not repeated in this guide to reduce repetition. Some organisms are covered in a habitat that they frequent less than another; however, they are described out of their primary environment to compare them with similar species.

Because this book is written for the casual and the serious beachcomber, it differs somewhat in general format and organization from the more traditional academic approach to field guides. Information, advice, or comments are first directed to the casual beachcomber, with technical terms avoided as often as possible. Information is also included for serious beachcombers who are interested in the examples of life they discover on the beach but don't intend to be marine biologists. They may want to collect samples for many reasons. For those not already familiar with the major groups of marine animals (e.g., cnidaria, mollusca) and plants (e.g., chlorophyta, anthophyta), Chapter 1 provides brief descriptions. For serious beachcombers some technical information is also included as an introduction to the classification system.

This guide has been structured for cover-to-cover reading by the student of the marine environment, and for spot reference by the casual beachcomber as the need arises. As a minimum, this Introduction should be followed by Chapter 1 (Classification of Marine Organisms) for descriptions of the major groups of marine animals and plants. Chapters 2 through 6 provide orientation to the five types of "beach" environment you may encounter, and may be read selectively, as required. Organisms mentioned in other chapters are listed by their scientific and common names. Also, each organism mentioned in the book is listed under its scientific name as well as its common name(s) in the index.

Contrary to popular belief, marine biologists don't really speak in Latin or Greek, but they do use those "neutral" languages to identify to all the biologists of the world exactly what they have found, observed, and studied. That is difficult to do in Portuguese or Balto-slavic. Thus, the organisms' scientific names are given for serious beachcombers. Actually, most plants and animals only have a scientific name—no common name. Also, common names for marine organisms vary from region to region, and a serious beachcomber wanting more information on an organism must use its scientific name for further study.

Chapter 1, dealing with marine animals and plants, includes a general introduction to the phyla and divisions (highest classification within a kingdom for animals and plants), and then (only for animals) is followed by information to identify the class in which an organism belongs. A few illustrations of the anatomy of major groups of animals will enhance your understanding of organisms within those groups. Further, the quick reference key on page 4 will direct you to the appropriate phylum and class of the animal (or division, if it is a plant) you have found.

Water temperature is the reason for restricting the study area to between Cape Cod and Cape Hatteras. The temperature of coastal waters north of Cape Cod and south of Cape Hatteras may vary from each other by as much as 30°F. The primary physical feature that influences the geographic range of a marine organism is water temperature.

As the warm water of the Gulf Stream flows up from the south, carrying 25 times the combined flow of all the earth's rivers, the current is deflected by Cape Hatteras, which projects 12 miles into the Atlantic, moving it offshore. North of the cape, water temperatures are cooled by the Virginia Current, an offshoot of the southbound Labrador Current, which slides under the warmer waters of the Gulf Stream. As the Gulf Stream continues northward, eddies of its warm water curl off and head toward shore, slightly warming the cold water of the Labrador Current. What are usually considered tropical species of organisms are often carried into the study area by those eddies. Shaped like a flexed arm, Cape Cod curls 25 miles into the Atlantic, further deflecting the Gulf Stream offshore. Off the coast of Cape Cod at Georges Bank, the Gulf Stream intersects the very cold Labrador Current and streams off to the northeast, while most of the Labrador Current continues southbound along the coast.

The Gulf Stream received its name from the misconception that its source was the Gulf of Mexico. Actually, the Gulf contributes very little water. The Equatorial Current flows through the passage between the Windward Islands into the Caribbean Sea. The current, flowing through the Yucatan Channel, has only one outlet, the Straits of Florida. Off Florida other currents, from the northern coast of Puerto Rico and east of the Bahamas add to the flow.

The water in the Gulf Stream is blue for the same reason that the sky is blue. The blue water is caused by scattering of sunlight by tiny particles suspended in the water. Short wavelength blue light scatters more effectively than light of longer wavelengths, such as red. Green or brown water is commonly seen near coasts. The pigments of microscopic floating plants (phytoplankton) often are responsible for the color. Also, the silt and sediment from beaches and outflows of large rivers turn the water a brown color.

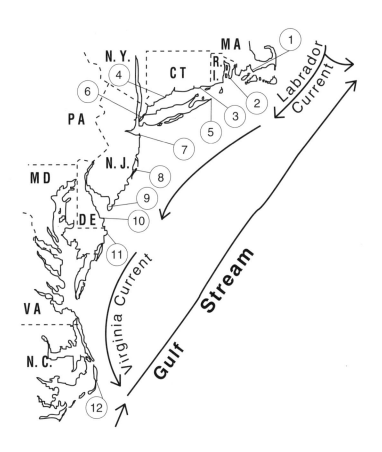

Water Temperatures* (°F)

Beach Locations	Jan	Feb	Mar	Apr	May	June	July	Aug	Sept	Oct	Nov	Dec
1. Woods Hole, MA	34	35	37	45	55	62	71	71	67	59	50	41
2. Newport, RI	37	36	37	46	54	62	68	70	66	60	52	44
3. New London, CT	37	37	40	49	56	64	70	71	68	59	52	42
4. Bridgeport, CT	39	37	40	48	58	67	72	76	72	63	55	45
5. Montauk, NY	36	35	38	43	52	60	68	70	67	59	56	43
6. The Battery, NY	38	36	41	47	57	65	71	73	70	60	53	43
7. Sandy Hook, NJ	37	36	40	46	55	61	69	72	67	59	51	43
8. Atlantic City, NJ	37	35	42	48	56	63	69	72	69	60	53	44
9. Cape May, NJ	37	37	42	49	59	68	72	73	71	60	52	42
10. Lewes, DE	37	36	41	51	60	67	72	75	71	61	52	44
11. Ocean City, MD	37	34	42	49	55	62	68	71	70	62	53	44
12. Cape Hatteras, NC	49	46	52	59	68	73	77	80	76	70	58	55

*The water temperatures were provided by the National Ocean Survey tide stations over an eight-year period.

The temperature of the water and its salinity affects the organisms you may find on various shores. Some species of marine organisms are found along the entire U.S. Eastern Seaboard. However, the southern range of many organisms stops at the bayside of Cape Cod, while the northern range of many organisms does not extend beyond Cape Hatteras. Most of those described in this book range throughout the study area or from Long Island northward and Long Island southward. However, a very few will be found only on the bayside of Cape Cod or at Cape Hatteras and nowhere else in the study area.

Four terms are used to classify marine animals found in the study area. Those living on rocks and firm sediments on the bottom are referred to as *epifauna* (plants are *epiflora*). Those buried in the substrate are *infauna;* they are usually found in soft sediments, such as sand and mud. Animals attached to the substrate are *sessile* (e.g., barnacles and mussels). Others, such as snails and crabs, that move about, are *motile*.

OBSERVING AND COLLECTING

A serious beachcomber should not be without a magnifying glass, 10-power or better, to observe small organisms. Binoculars are useful for identifying shore birds and enjoying their antics while searching for food. Most will not let you approach too closely, others fly over the water in search of prey. A dip net to sweep through the shallow water of bays and estuaries is also needed. A trowel or shovel (the small folding entrenching tool sold in army surplus stores) and a sieve or sifting box can be used to collect burrowing animals from mud flats. Also, rubber boots would make the mud flats and salt marsh more navigable. Ziploc-type plastic storage bags and a wide-mouth plastic (*never glass* for obvious reasons) container to hold the bags should be included. A pencil and notepad are needed to record data (location, time, conditions, etc.) relating to the collected specimens.

Beachcombing is a leisurely, stressless adventure in what is probably the most beautiful classroom in the world. You can find an entire world in microcosm along a single mile of beach. Unfortunately, you can also encounter human jetsam such as used syringes, black balls of congealed petroleum, wood with nails, and glass—examples of our desecration of the water and the beaches.

Be especially careful if you explore jetties (man-made rock protuberances into the water from the beach), for they are mainly intertidal (between the sea and land) and the rocks are often slippery. The jetties are seldom totally safe, but that doesn't mean you should ignore them. They are an excellent substrate for the attachment of marine organisms. But you should know the risks involved. Most seaweeds (algae) attached to rocks are very slippery. Many beachcombers who have fallen among them have suffered broken limbs and more serious injuries. An internationally known American botanist fell while collecting seaweeds on a jetty, struck his head on a rock, slipped into the water, and drowned. I suggest wearing gripping, rubber-soled shoes (or sneakers) for exploring rocky beaches and jetties.

Conservation and Regulations

If you are beachcombing at a national seashore, you are not allowed to remove living organisms. After carefully collecting and observing the animals gently place them back where they were found. A notebook, pencil, and camera, with a close-up lens, will furnish a record and a means of "collecting" specimens. National Seashores were established by the National Park Service to preserve the barrier islands and their flora and fauna. Because of such protection, our descendants will be able to enjoy the unspoiled beaches for generations. The five national seashores in or near the study area are at Cape Cod (Massachusetts), Fire Island (Long Island, New York), Assateague Island (Maryland and Virginia), and Cape Hatteras and Cape Lookout (North Carolina).

Most states and towns have regulations restricting the collection of animals and set size limits for certain species. For example the blue crab must be of a specified size before it can be collected. However, an undersized crab is better suited for a small aquarium. Maryland prohibits the taking of blue crabs in November, and Virginia has a breeding sanctuary for crabs between Hampton Roads and the Atlantic Ocean. For shellfishing there are size limits, closed seasons, and closed areas. Check with the state or town conservation office for local restrictions before removing organisms from the beach.

If you are not at a protected seashore and the organisms are to be taken back for laboratory observation or to place them in an aquarium in your home, fill plastic bags with water from the collection site before putting animals or plants into them. A water-filled plastic bag is also recommended when making field observations, even if the organism is to be released back into the water. The delicate filamentous algae will spread out in the water and small animals like the beach hopper and the glass shrimp will swim about.

If the organisms are to be removed, a cooler with ice, in which to place the plastic bags, might be necessary. Removing organisms from cold or cool water and leaving them exposed to changes in temperature could be fatal for them.

Be observant and patient; otherwise you'll think the landscape is barren. Some things, such as sand, pebbles, shells, and seaweed are easy to find and observe. Others like the burrowing sand flea and mole crab are more elusive. The fun part of beachcombing is that you never know what you might see. Beachcombing can last all year long, although it might not be as pleasant during winter months, but it is time well spent.

I wish you good reading, good learning, good collecting, good fun in the fascinating classroom of Atlantic beaches. Beachcombing is an activity of a lifetime. The more you know, the more you appreciate.

Hank Keatts
East Moriches, New York

About the Author

Henry Keatts is a professor of biology and oceanography, Suffolk Community College, Long Island, NY. In addition to being widely published in his field Keatts is the author of *New England's Legacy of Shipwrecks* and *Field Guide to Sunken U-boats,* both published by the American Merchant Marine Museum Press (United States Merchant Marine Academy), and *Guide to Shipwreck Diving: New York and New Jersey* published by Pisces Books. He is co-author of the *Dive into History* series (*U-boats, Warships, and U.S. Submarines*) also published by Pisces Books. Keatts writes a column "History Submerged" for *Discover Diving* magazine. He is a "Fellow" of the prestigious Explorers Club (Manhattan, NY), an associate member of the Boston Sea Rovers and an honorary member of the Gillmen Club (Hartford, CT) and the Adirondack Underwater Explorers (Saratoga Springs, NY). Professor Keatts is president of the American Society of Oceanographers and editor of the Society's *Journal.*

Over the past 14 years he has presented over 130 slide/lecture presentations on marine biology and maritime history to such varied groups as the U.S. Naval War College, service organizations (Kiwanis, Lions, and Rotary), historical societies, Museum of Natural History (Manhattan), Explorers Club (Manhattan), dive clubs, church groups, and colleges and universities (U.S. Merchant Marine Academy, Harvard, Temple, Brown and George Washington to name a notable few).

1

Classification of Marine Organisms

To him who in the love of Nature holds
Communion with her visible forms, she speaks
A various language.

—William Cullen Bryant

The marine environment is a very large, complex system. The variety of marine organisms and the profusion of common names applied to them create a massive identification problem. Common names vary from one country to another and from region to region within a country; but the scientific name is universal. In New England, a fish may be referred to by one name and in the Gulf States by another. Different common names are even used in the same geographic area. For example, in the northeast the anglerfish has several other common names including goosefish, headfish, and monkfish (Figure 1-1).

The attendant confusion of common names was responsible for Carolus Linnaeus, a Swedish botanist, introducing a binomial nomenclature system in 1758 to facilitate identification. It established the scientific names that are used today in all technical journals, regardless of

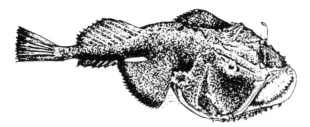

Figure 1-1. *Lophius americanus* has many common names: anglerfish, goosefish, headfish, and monkfish.

the researcher's nationality. Each scientific name includes a capitalized generic (genus) name, followed by an uncapitalized specific (species) name. Both genus and species are italicized or underlined. For example, the scientific name of the anglerfish referred to above is *Lophius* (genus) *americanus* (species). Most animals and plants have only scientific names.

The use of Latin and Greek names makes biology in general, and taxonomy in particular, seem unnecessarily formidable. When the system was created, it was decided that the scientific names used for classification should be written in a universally familiar language. English and French were both considered, but were rejected because neither was acceptable to the other country. Therefore, two classical languages were proposed—Latin and Greek. In this guide L. indicates Latin, Gr. Greek. Although the terms are foreign to most Americans, they are a necessary tool.

Throughout the book, common names are given whenever possible, but in all cases the scientific names. Occasionally another well established generic or specific name is in parenthesis or brackets.

The major categories (taxa) of the taxonomic hierarchy now used are:

Kingdom ⟶ Phylum ⟶ Class ⟶
(botanists often use Division)
Order ⟶ Family ⟶ Genus ⟶ Species

This biological classification system is hierarchical, with "kingdom" being the broadest and most inclusive category and "species" the least inclusive. An organism is assigned to one small set of very similar individuals (species), and that set is grouped with other sets to form a larger set of individuals with some similarities (genus), and so on through family, order, class, phylum or division, and finally kingdom.

In addition, subdivisions of the categories (subphylum, subclass, etc.) may be applied to the organisms, as required for full identification. In dividing phyla into classes, orders, etc., some single characteristic or combination of characteristics of each group of organisms excludes all others from the group. This system provides a concise, orderly method by which the more than three million living organisms can be named and classified.

Table 1-1 illustrates the classification hierarchy by comparing the taxonomic classification of a human and a blue crab.

Current classification schemes are based on theoretical evolutionary relationships, and individuals interpret relationships among organisms differently. Scientists often do not agree on how many phyla or other taxa there are, and different texts list different numbers. International commissions on nomenclature meet from time to time to formulate rules and make taxonomic decisions.

Table 1-1
Taxonomic Comparison

Human	Blue Crab
Kingdom Animalia—multicellular animals	Kingdom Animalia—multicellular animals
Phylum Chordata—notocord, nerve cord	Phylum Arthropoda—jointed appendages
Subphylum Vertebrata—vertebrae (backbones)	
Class Mammalia—mammary glands	Class Crustacea—crustlike body covering
	Subclass Malacostraca—soft-shelled
Order Primates—superior nervous system, nails on digits	Order Decapoda—ten legs
Family Hominidae—no tail or cheek pouches	Suborder Brachyura—short tail
Genus *Homo*—manlike	Family Portunidae—from Portunus, god of the harbor
	Genus *Callinectes*—from two words meaning "beautiful" and "swimmer"
Species *sapiens*—wise, knowing	Species *sapidus*—palatable

QUICK REFERENCE KEY

The extensive literature of marine biology provides the layman with the means to identify any marine organism, as long as he or she has unlimited time and patience. Table 1-2 provides a series of steps to permit a casual beachcomber to identify forms of marine life relatively quickly based on observed physical characteristics.

The first step is to determine whether or not the observed organism is encased in a shell (not an exoskeleton like that of a crab or lobster). If it is, attention is directed to choice "A" to determine whether the shell is of one, two, or eight parts. If there is no shell, the search is directed to choice "D" (body feel). If it is soft, reference is directed to letter "F" (body appearance). The search continues in that manner until a phylum, and usually a class, is designated. Refer to the index for descriptions of phyla and classes.

Most, but not all, organisms found in the study area can be placed in a phylum or class by following the quick reference key. Table 1-3 shows a similar, simpler key for the quick classification by division of major groups of marine plants.

You may have difficulty in distinguishing between an internal skeleton, such as is found in a sea star (starfish, Phylum Echinodermata) and the external skeleton (exoskeleton) of a crab or lobster, Phylum Arthropoda. However, usually members of both phyla have hard bodies, so the beachcomber is directed to the letter "G" of Table 1-2.

MARINE ANIMALS

All animal phyla have marine representatives. Some are even entirely marine. Not all of the animal phyla are presented in this guide, only those that are most frequently encountered in the study area by a beachcomber.

Phylum Porifera (po-rif′era) (L. *porus*, pore + *ferre*, to bear). Porifera means "pore bearer," from the numerous tiny holes that perforate their bodies. These animals are called sponges; it is one of the few groups with a widely accepted common name. The sponges are one of

Table 1-2
Key for Quick Reference to Major Groups of Marine Animals

		Go To
If *SHELL* is	present	A
	absent	D
A. If *SHELL* has	one part	B
	two parts—Phylum Mollusca, Class Bivalvia	
	eight parts—Phylum Mollusca, Class Amphineura	
B. If *SHELL* is	tube-like—Phylum Annelida, Class Polychaeta	
	snail-like—Phylum Mollusca, Class Gastropoda	
	volcano-shaped	C
C. If *SHELL* has	opening at top of shell	E
	opening absent—Phylum Mollusca, Class Gastropoda (limpet)	
D. If *BODY* is	soft	F
	hard	G
E. If *SHELL* has	fused plates—Phylum Arthropoda, Class Crustacea (barnacle)	
	one part—Phylum Mollusca, Class Gastropoda (keyhole limpet)	
F. If *BODY* is/has	gelatinous	H
	spongy	I
	slug-like—Phylum Mollusca, Class Gastropoda	
	segmented—Phylum Annelida, Class Polychaeta	
	feathers—Phylum Chordata, Class Aves	
	feeding tentacles & tube feet—Phylum Echinodermata, Class Holothuroidea	
G. If *SCALES*	....	J
If SPINES are	present—Phylum Echinodermata	
	absent	K
ALMOST MICROSCOPIC usually encrusting—Phylum Bryozoa		
H. If *TENTACLES* are present		L
	absent—Phylum Ctenophora, Class Tentaculata	
I. If *SQUIRTS*	water when squeezed—Phylum Chordata, Subphylum Urochordata	
	does not squirt water—Phylum Porifera	

continued on next page

Table 1-2 Continued

J. FISH	Phylum Chordata
SNAKE or *TURTLE*	Phylum Chordata, Class Reptilia
K. If *SKELETON* is	external (crab- or shrimp-like carapace)—Phylum Arthropoda
	internal—Phylum Chordata
L. If *EYES* are	present—Phylum Mollusca, Class Cephalopoda
	absent—Phylum Cnidaria

Table 1-3
Key for Quick Reference to Major Groups of Marine Plants

If *ROOTS* and *FLOWERS* are	present	.... Division Anthophyta
	absent	.... **A**
A. If *COLOR* is	green	.... Division Chlorophyta
	brown	.... Division Phaeophyta
	red	.... Division Rhodophyta

the simplest forms of multicellular animals. Except for simple epithelium (surface tissue), they do not possess tissues (specialized, coordinated group of cells). For that reason, sponges are considered to be on the cellular level of organization instead of the tissue and organ level. However, sponges do demonstrate division of labor; their cells are specialized to perform the different functions of supporting the body, gathering food, and reproduction, which can be asexual or sexual. They do not react when touched because they lack muscle or nervous tissue.

Sponges are filter feeders; they have specialized cells, called choanocytes or collar cells, with flagella (long hair-like projections). Those cells line the spongocoel (interior cavity) and the rapid beating of their flagella pulls water, heavily laden with microscopic food, through the many tiny perforations in their body. As the water flows over the choanocytes, their flagella propels the food (algae, bacteria, etc.) down toward the cell body, which engulfs it, forming food vacuoles. Digestion is intracellular. Cells (amoebocytes) in a gelatinous middle layer transport and store food particles. The water leaves the sponge's body either directly through one large opening or through several large openings, called oscula. (Color Plates 2c and 3b).

Aristotle in his *Historia Animalium* considered sponges to be intermediate between plants and animals. But Carolus Linnaeus and his contemporaries in the 18th century included them in the plant kingdom. John Ellis, a great British naturalist and a Fellow of the Royal Society, persuaded Linnaeus to place sponges in the animal kingdom. Controversy continued until 1825, when Dr. R.E. Grant of Edinburgh observed under the microscope "this living fountain vomiting forth, from a circular cavity, an impetuous torrent." He proved that sponges possessed the animal property of feeding on organic matter and not on inorganic matter as do plants.

The word sponge is usually associated with household cleaning chores or washing the car. However, most sponges purchased in stores are synthetic. When it is the marine animal, it is actually the sponge's skeleton, a network of flexible spongin (protein) fibers. Commercial sponges are found in warmer waters than those covered by this guide. They are collected, and left to dry in the sun so cellular decomposition will occur, leaving the skeleton.

Other types of sponges have skeletons of hard mineralized spicules, which are either calcareous (chalklike) or siliceous (glasslike) (Figure 1-2). The size and shape of the spicules differ in each species, and the classification of sponges is based on the variety of those supportive

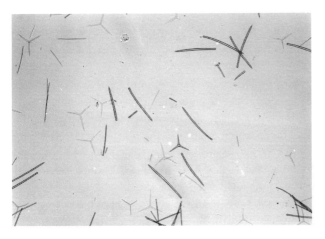

Figure 1-2. An assortment of sponge spicules, magnified 40 times.

structures. The sponge body consists of three basic strata: an outer layer of epithelium, an inner layer of choanocytes, lining the spongocoel, and a gelatinous middle layer (mesoglea) containing amoebocytes and spicules (Figure 1-3).

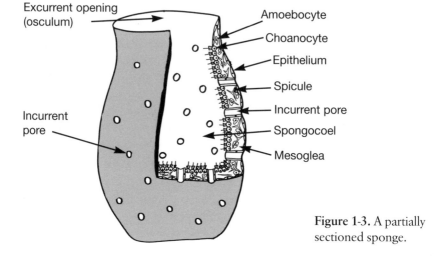

Figure 1-3. A partially sectioned sponge.

Sponges reproduce asexually by budding or fragmentation, and sexually when choanocytes or amoebocytes function as gametes.

Living sponges have a gelatinous texture, quite different from those used around the house. They may be yellow, red, blue, or green, the latter caused by a green unicellular alga living in the sponge's body. They are usually found attached to hard substrates such as rocks and pilings; many animals, such as crabs, feed on sponges (Figures 1-4 and 1-5).

Phylum Cnidaria (ny-dar'e-a) (Gr. *knide*, nettle + L. *aria*, like are connected with). Some prefer the phylum name Coelenterata (se-len'te-ra'ta) (Gr. *koilos*, hollow + *enteron*, gut + *ata*, characterized by) for this group of animals. This phylum contains the familiar jellyfish, sea anemones, and corals. These sac-like (cylindrical body) animals are thought to be the first to develop several types of tissues, (epithelial, nervous, digestive, and muscular), and hence reach the tissue level of organization. The cnidarian body wall consists of three basic strata: an outer layer of epidermis (epithelium); an inner layer of gastrodermis,

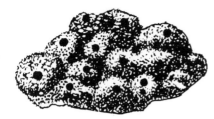

Figure 1-4. Encrusting sponge.

Figure 1-5. Erect sponge.

lining the gastrovascular cavity (enteron); and a middle layer called mesoglea. The mesoglea ranges from a thin membrane to a thick, clear, gelatinous, or jelly-like mass characteristic of jellyfish. Although cnidarians have true tissues, they do not possess organs.

Tentacles, which usually surround the mouth, possess special cells called cnidoblasts, with stinging structures called nematocysts. The cell has a projecting triggerlike cnidocil, which is stimulated by tactile or chemical stimuli. Also, nervous impulses will activate the trigger. When the cnidocil is stimulated, the coiled nematocyst is extended with explosive force, penetrating the prey and releasing a toxin (Figure 1-6). Some nematocysts have a sticky substance on the surface and wrap around the prey, entangling it and depositing venom. After the nematocyst is discharged, it is released by the cell and adjacent cells produce new ones. Those epithelial cells, cnidoblasts, used to capture prey and for defense, account for the fact that the phylum name is cnidaria.

After capturing prey, the tentacles are used for ingestion of the food through the mouth into the gastrovascular cavity. Gland cells in the gastrodermis secrete enzymes and digestion (extracellular) begins within the cavity. After initial disruption of the food, small particles are taken into other cells by pseudopodia (projections of the protoplasm), where intracellular digestion occurs. Digested food can be stored in the gastrodermis

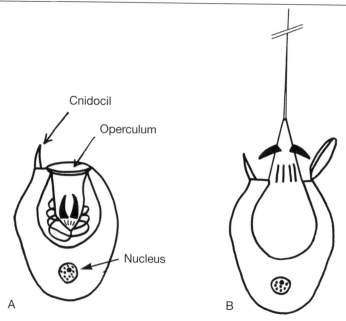

Cnidocil

Operculum

Nucleus

A   B

Figure 1-6. A cnidoblast, before (A) and after (B) discharging the nematocyst.

and passed by diffusion to the other tissues as needed. Because there is only one opening, egestion of non-digested material is also through the mouth. The one opening functions as both mouth and anus.

Cnidarians may exist either individually or in colonies. Two distinct body types are recognized: the sessile (living attached) polyp, usually representing the asexual (normally by budding) phase of the life cycle, and the free-swimming medusa, the sexual phase. Polyps are tubular with a mouth at one end, surrounded by tenacles. The opposite end is closed, and forms a flattened attachment organ, the pedal disk. The cylindrical body column between the two ends is highly contractile. When contraction occurs, the polyp becomes much smaller. Medusae are delicate transparent animals with bell- or umbrella-shaped bodies, such as a jellyfish. Around the periphery, or edge of the umbrella, tentacles surround a mouth, centrally located on the undersurface. Most species do not exhibit both body types, but in some species, each generation gives rise to the alternate type, referred to as alternation of generations. This phylum includes three classes. Classification is primarily based on the degree of dominance or suppression of one of the body types, and secondarily on other characteristics.

One of the great discoveries of the 19th century was that some hydrozoans and scyphozoans exhibit alternation of generations. The remarkable life cycle was first described in 1829 by the Norwegian pastor Michael Sras, who later became professor of zoology at Christiania in Oslo.

Class Hydroza (hy-dro-zo'a) (Gr. *hydra*, water serpent + *zoon*, animal). Members of this class are small and inconspicuous and are referred to as hydrozoans or hydroids. Typically, both polyp and medusa are present in the life cycle, although one type may be suppressed. The jellyfishlike medusa, when present, is dioecious (either male or female). The polyps are often mistaken for plants. Different types of polyps may be present in a single colony. Some polyps are specialized for gathering food with their tentacles and others are for reproduction, producing medusae (Figures 1-7, 4-2, and 4-3). Hydroids feed primarily on zooplankton (microscopic animals) and are found attached to hard substrates such as rocks and pilings, and on eel grass and seaweeds.

Class Scyphozoa (sy-fo-zo'a) (Gr. *syphos*, cup + *zoon*, animal). Members of this class, commonly called jellyfish, do not form

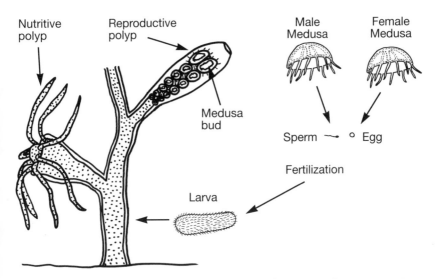

Figure 1-7. The colonial hydroid *Obelia* exhibits alternation of generations in its life cycle, and has both feeding and reproductive polyps.

colonies. The medusa is the dominant body form, and the polyp stage is reduced or absent. Sensory pits are typically found at the periphery of the umbrella. These are statocysts for equilibrium and ocelli, small simple eyes that determine light intensity but do not form images. The large amount of gelatinous mesoglea, the body's middle layer, makes the animal more buoyant and it floats easily. The thick jellylike layer gives rise to the inappropriate name jellyfish (Figure 1-8, and Color Plate 4a). Scyphozoans are not related to fishes and their body structure is not close to the complexity of that of vertebrates such as fishes.

Figure 1-8. Jellyfish.

Although influenced by waves and currents, medusae are good swimmers; it is fun to watch the graceful undulations of the umbrella propelling the animal through the water. However, the trailing tentacles of some species should be avoided because they contain stinging cells that are used to catch small fishes and crustaceans. However, only about a dozen of the more than 500 species of jellyfish worldwide are venomous, and of those, only a few are hazardous to humans. Like the hydroid medusae, sexes are separate. The reproductive organs can usually be seen through the umbrella. Small scyphomedusae can be distinguished from hydromedusae by the umbrella margin of the latter that projects inward, producing a shelf called the velum.

Class Anthozoa (an-tho-zo'a) (Gr. *anthos*, flower + *zoon*, animal). Anthozoans are often called flower animals. These relatively stationary animals are often mistaken for plants because the many tentacles of the polyps give them a flowerlike appearance. All anthozoans are polyps; with no medusa stage there is no alternation of generations in this class.

They differ considerably from hydrozoan and scyphozoan polyps. A tubular pharynx extends from the mouth to more than halfway into the gastrovascular cavity. The gastrovascular cavity is divided by radially positioned septa (partitions) into compartments, which allow anthozoan polyps to be larger and heavier than hydrozoans or scyphozoans. In addition to supporting the body column, the septa also increase the efficiency in digestion.

Sea anemones and corals are familiar representatives of this class (Figure 1-9, and Color Plates 4b, 4c, and 4d). Simple unbranched tentacles are characteristic. Most anthozoans live attached to hard substrates such as rocks and pilings, but a few burrow in sand or mud. Generally they feed on plankton and small invertebrates.

Figure 1-9. Sea anemones, with tentacles contracted and extended.

Phylum Ctenophora (te-nof'o-ra) (Gr. *ktenos*, comb + *phora*, bearing). Ctenophores, resembling jellyfish in appearance, are close relatives of cnidarians and are commonly found in surface waters. They have a thick middle body layer comparable to the cnidarian mesoglea, are spherical or ovid in shape, and are often mistaken for jellyfish.

However, major differences separate the two groups of animals. Ctenophores do not have an alternative body type such as a polyp, nor do they have cnidoblasts, except in one species, which is not found on the East Coast. Eight rows of characteristic ciliated (cilia are hairlike projections from cells) bands, called comb rows, have given the phylum its name. The comb rows are used in locomotion instead of using muscular contractions of the umbrella like cnidarian medusae. Scattered muscle cells are present along with the epithelial and nervous tissue. Ctenophores, like cnidarians, are still on the tissue level of organization; they do not possess organ systems.

Ctenophores feed on plankton they capture with small adhesive cells (colloblasts) on their tentacles, which are then wiped into the mouth. Statocysts (for equilibrium) are their only sense organs. They only reproduce sexually. When disturbed, ctenophores produce luminescence, a beautiful phenomenon that can easily be observed by watching the wake of a boat at night.

Class Tentaculata (ten-tak-yu-la′ta) (L. *tentaculum,* feeler + *ata,* group). This class includes the species of ctenophores that possess tentacles, such as the sea gooseberry (*Pleurobranchia pileus*) (Figure 4-15, and Color Plate 5a).

Phylum Mollusca (mol-lus′ka) (L. *molluscus,* soft). The name means "soft bodied." The mollusks are the most diverse group of marine animals and the most conspicuous invertebrates. The phylum includes snails, mussels, oysters, clams, squid, and octopi. This phylum is very important commercially.

Their tissues are well developed, with several types of tissue cells combined to form organs for specific functions. Thus, they are organized on the organ-system level with a digestive system, excretory system, respiratory system, nervous system, and reproductive system. Well-developed sensory organs of touch, smell, taste, vision, and equilibrium are often present.

The animals usually have hard calcareous shells, generally of two types: a single coiled shell carried on the back (as in snails) or two "valves" enclosing the body (as in clams). The shell functions as an exoskeleton, a site for muscle attachment. Basically, the body consists of a head, a large foot, and a visceral mass containing most of the organs. The dorsal body wall forms a fleshy mantle that covers the body and secretes the shell when present. Calcium carbonate is absorbed from the water, then is secreted by glands within the mantle to produce and enlarge the shell. The shell is composed of three layers; the outer two often show concentric growth rings because the material is not laid down continuously. Other ridges and projections on the shell are determined by the folds of the mantle's edge. The scallop's mantle edge is waved or frilled and produces fanlike ribs. Pigment cells in the mantle produce the colorful patterns. Often the shell has a protective covering of organic material, called a periostracum, which is also secreted by the mantle. It may be thin, appearing transparent or pale-colored, or heavy and ornamented with hairs or bristles. The periostracum often flakes off of shells that have been out of the water for a long period.

Mollusks are popular, not just to eat but to collect. One danger often overlooked is over-collecting live specimens by individuals or classes. Some species have been brought almost to extinction by shell collectors.

A superficial survey of the classes of mollusks seems to indicate a dissimilar group of animals. Snails, oysters, and octopi appear to have little structural similarity. Yet all mollusks have the same fundamental body plan.

Class Amphineura (am'fi-neu'ra) (Gr. *ampli*, on both sides + *neuron*, nerve)—also known as class polyplacophora. The most distinctive characteristic is a flattened body with a shell consisting of eight overlapping plates on the dorsal surface (Figure 1-10). Because of these plates and their articulation with one another, these animals are able to roll up in a ball when they are dislodged from the substrate. Members of this class frequently found in the study area are called "chitons," a Greek term meaning tunic. A broad flat foot occupies most of the ventral (lower) surface. It functions in adhesion as well as locomotion. Most chitons feed on algae they scrape from the surface of rocks and other hard objects. They possess a rasping tonguelike organ called a radula, which is projected from the mouth to scrape off the algae.

Figure 1-10. Chiton.

Class Gastropoda (gas-trop'o-da) (Gr. *gaster*, belly + *podos*, foot). This is the largest class of mollusks. Most species possess shells, but some entirely lack one. The shell, when there is one, is univalved (one piece) and varies considerably in shape (Figure 1-11). We've all been told as children that we can hear the sound of the sea in a large gastropod shell. Actually, the sound is the echo of the pulse in our heads or the reverberations of wind or barometric pressure.

Gastropods undergo a process called "torsion" very early in embryonic development. The visceral mass is twisted 180° allowing the head end of the animal to be drawn into the protection of the mantle cavity within the shell, before the less vunerable foot. Often a calcareous or

Figure 1-11. Gastropod shells vary considerably in shape and size.

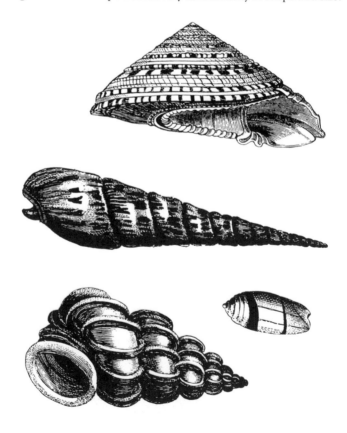

horny disc (known as an operculum) is attached to the tough foot and it closes the shell opening when the foot is withdrawn (Figure 1-12). Before torsion occurred, the head could not be rapidly retracted into the mantle cavity because the foot was in the way. Twisting of the mantle cavity, from posterior to anterior, permits the sensitive head to be retracted first and then the tougher foot. The process makes it easier to escape many predators. In gastropods without shells, such as nudibranchs (sea slugs), detorsion follows torsion, partially restoring the original bilateral symmetry (Figure 1-13).

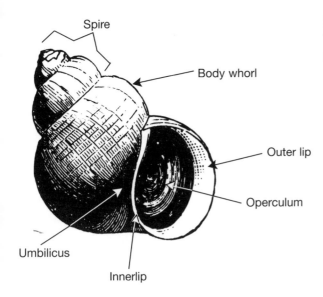

Figure 1-12. Terminology of univalve shells.

Spire

Body whorl

Outer lip

Operculum

Umbilicus

Innerlip

Figure 1-13. Nudibranchs (sea slugs) are gastropods without shells.

In most gastropods the rasping tonguelike radula, covered with rows of sharp teeth, is used for feeding (Figure 3-14). Feeding habits vary among the species. Some are herbivorous and scrape algae off rocks. Others are carnivorous and drill holes through the shells of other mollusks. Many are scavengers.

Locomotion is accomplished by muscular contractions of the extended foot. One or two pairs of tentacles bear eyes and tactile and chemoreceptor cells. The eyes are simply photoreceptors in many species, but some do possess image-forming eyes. Some species have both male and female reproductive organs, others have only one type. Some have copulatory organs, some do not.

Class Bivalvia (bi′val-via) (L. *bi*, double + *valvae*, folding doors)— also known as class Pelecypoda (pel-e-sip′oda) (Gr. *pelekus*, hatchet + *podos*, foot). Members of this class are usually referred to as bivalves. They are laterally compressed, and possess a shell with two parts (valves) hinged dorsally by an elastic ligament made of protein. Small, interlocking teeth near the hinge keep the valves in position. Each valve has an elevated area, the umbo, which is the oldest part of the shell. Growth rings are produced as the mantle secretes successive layers of shell. The rings radiate out from the umbo (Figures 1-14 and 1-15).

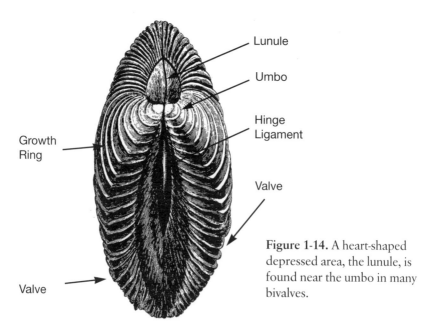

Lunule

Umbo

Hinge Ligament

Growth Ring

Valve

Valve

Figure 1-14. A heart-shaped depressed area, the lunule, is found near the umbo in many bivalves.

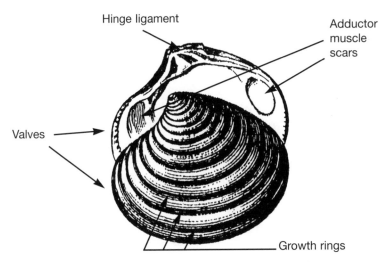

Hinge ligament

Adductor muscle scars

Valves

Growth rings

Figure 1-15. Terminology of bivalve shells.

Pearls are formed when a foreign object, such as a sand grain, lodges between the mantle and the shell, and is surrounded by concentric layers of shell to reduce the irritation. Most bivalves can produce pearls, but the most valued are found in pearl oysters (not found in the study area). Oysters usually produce only one to three pearls. However, *The Guinness Book of Animal Facts and Feats* recorded one specimen from the coast of China that contained 1,176 pearls.

The two valves (shells) of most bivalves are held together by the contraction of two large muscles, the anterior and posterior adductors. Each contains two types of muscle tissue, striated fibers for fast contraction and smooth fibers for sustained contraction. When the muscles relax, the valves are drawn open by the elasticity of the dorsal hinge ligament that connects them. Bivalves such as oysters and scallops have only one adductor muscle, more centrally located. Conspicuous scars are present on bivalve shells where adductor muscles were attached (Figure 1-15). Also, where the mantle attaches near the valve edge, a scar, called the pallial line, is clearly seen (Figure 1-18).

Most bivalves attach to a solid object, or burrow into the soft bottom. Because they are sessile or burrowing animals, the head is greatly reduced in size and the sensory organs are redistributed to other parts. Most have enormously enlarged, mucus-covered gills for filter feeding and respiration (Figure 1-16). Hairlike cilia pass food-laden mucus to the mouth for ingestion. In burrowing bivalves the posterior end of the mantle produces two tubular siphons that extend to the surface of the sand or mud (Figure 1-17). One is an incurrent siphon to bring water to the gills, the other, an excurrent siphon, expels the wastewater. When observing an empty shell, an indentation in the pallial line, called the pallial sinus, indicates where the siphons were located (Figure 1-18).

The gills of bivalves living in polluted water can contain bacteria that cause diseases in humans, especially if eaten raw. When areas are restricted from shellfishing, it is for a good reason.

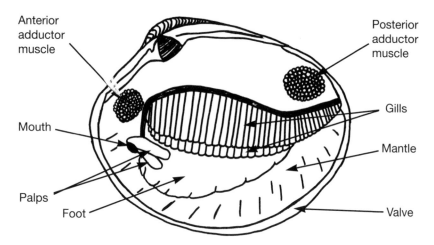

Figure 1-16. Anatomy of a clam, with one valve removed. Food trapped in mucus on the gills is passed to the mouth by the palps.

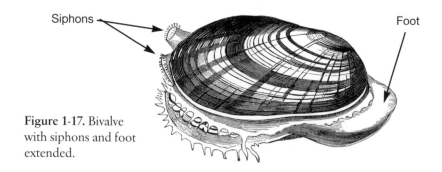

Figure 1-17. Bivalve with siphons and foot extended.

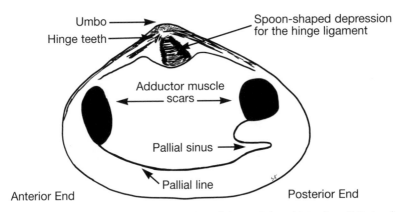

Figure 1-18. An interior view of one valve of the surf clam (*Spisula solidissima*) exhibits typical bivalve shell markings.

Bivalves such as clams and oysters shed their gametes (sex cells) into the water and fertilization is somewhat chancy. As many as 24 million clam eggs will be released in one spawning, 100 million eggs by an oyster.

Commercial industries developed around clams, oysters, scallops, and mussels; they make a significant contribution to the economy of many waterfront communities.

Class Cephalopoda (sef′a-lop′o-da) (Gr. *kephale*, head + *podos*, foot). This class contains the familiar squid and octopi, which are far more advanced and complex than other mollusks. These carnivorous predators have a large, well developed head and a highly developed nervous system, with eyes remarkably similar to human eyes and capable of forming images. The structure of the eye is similar to that of vertebrates. An iris diaphragm controls the amount of light entering the eye. A ciliary muscle controls the shape of the lens, allowing for depth perception. Even external muscles enable limited movement of the eye. In contrast with other mollusks, the external form of the cephalopod has little or nothing in common with the other classes. The foot that is common among mollusks and is used for creeping and swimming, has been replaced by powerful arms and tentacles. A strong pair of beaklike jaws, in addition to a radula, enable cephalopods to bite, and tear large pieces of tissue from their prey. The radula is used in a tonguelike action to pull the pieces of tissue into the mouth.

Cephalopods possess chromatophores (pigment cells) that allow them to change color and blend in with their background. A means of jet propulsion is used in locomotion by most cephalopods. Water passes into the mantle cavity and is forcefully ejected through a ventral tubular funnel (siphon). The stream of water propels the animal in the opposite direction. The funnel can be pointed in different directions and the force of water being expelled is controlled, resulting in beautiful swimming movements, especially in the squid.

Squid have fins on the end of the mantle that act as stabilizers. A squid has eight arms and two tentacles that are twice as long as the arms (Figure 1-19). The ends of the tentacles are spoon-shaped and covered with suckers. Tentacles catch prey and the arms, lined with suckers, hold the prey while it is being eaten. Squid along the U.S. coast have a thin, translucent internal shell called the pen, embedded in the mantle. Octopi have eight arms and do not have fins, tentacles, or shells (Figure 1-20).

Figure 1-19. Squid.

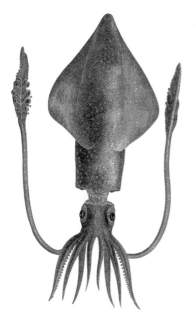

Figure 1-20. Octopus.

Most cephalopods are dioecious (separate sexes) and one arm of the male is modified with a scooplike depression to insert sperm into the mantle cavity of the female. Cephalopods range in size from tiny planktonic forms to giant squids that grow to almost 100 feet in length.

Like many other mollusks, cephalopods are of economic importance, not only as bait, but also as food (especially in the Orient and Europe). Cephalopods are low in fat and an excellent source of protein. Americans have been reluctant to eat squid, but consumption is on the rise. Octopus and squid are available in sushi bars; calamari (squid) is offered by Italian restaurants; deep-fried squid rings and strips are occasionally on the menus of other restaurants. The mantle, arms, and tentacles are edible.

Phylum Annelida (an-nel'i-da) (L. *annellus*, little ring + *ida*, suffix). Annelids are segmented worms, and are the most conspicuous of seashore worms. As the name implies, the most distinguishing characteristic of the phylum is the division of the body into similar segments called metameres. The introduction of metamerism, or segmentation, is the greatest advancement of this phylum; it lays the foundation for the more highly specialized metamerism of the arthropods. Metamerism is reflected externally as well as internally in all of the body organs, except for the digestive system (although it too extends through each metamere).

The digestive tube is located within a body cavity, called the coelom, which provides space for the development of the organ systems. All of the organ systems found in higher forms of animals such as humans, with the exception of the skeletal, respiratory, and probably endocrine, are present in the annelids.

Class Polychaeta (pol'e-ke'ta) (Gr. *polys*, many + *chaite*, long hair). Polychaetes are commonly called bristle worms. A distinct specialized head region, usually with differentiated sensory appendages such as tentacles, palps, and eyes, is present. Most body segments bear fleshy, branched, limblike appendages called parapodia that are the most distinguishing feature of this class. Many chitinous bristles, or setae, project from the parapodial divisions, giving the name to this class and strengthening the parapodia (Figure 1-21). The setae assume various shapes that are important in identifying polychaetes. Parapodia are

Figure 1-21. Polychaete annelid.

Parapodium with setae

used by polychaetes for locomotion, anchoring in tubes and burrows, and as the chief respiratory organs.

Most polychaetes in the study area covered by this guide are found in mud and sand. Thus, a sand screen is helpful for collecting these animals. Place a shovelful of mud and sand on the screen and wash with water. The worms and their tubes, if they are tube dwellers, will be left on the screen.

Phylum Arthropoda (ar-throp'o-da) (Gr. *anthros*, jointed + *pous*, foot). Approximately 75% of all animals are arthropods, numbering nearly a million species. Members of the phylum include such diverse animals as barnacles and crabs. Arthropods derive their name from their jointed appendages that bend only at the joints. The appendages of other animal phylas, such as the arms of an octopus and the parapodia of some annelids also bend, but throughout their length.

Only one other group of animals has a skeleton that is movable only at the joints, the vertebrates, but their skeletons are internal—arthropods' skeletons are external. Their unique body covering, or skeleton, is created by the secretion of epithelial cells. Chitin, a polymer, is the principal constituent of the exoskeleton. In many arthropods, especially the marine species, the skeleton also contains mineral salts such as calcium carbonate and calcium phosphate, making it thicker and more resistant to mechanical injury.

The armorlike exoskeleton is useful for support of soft body tissues, and also as a protective device. But, an external skeleton also presents problems. If the animal's body was entirely enclosed by a single skeletal structure, movement would be prohibited. Instead, it is composed of separate segments that are joined by a thin, flexible joint of pure chitin. Muscles attach interiorly to the two adjacent parts of the skeleton. When the muscle contracts, the joint bends, the limb moves, and the animal can walk. Marine arthropods attain a greater size than terrestrial forms because water helps to support the weight of their exoskeleton.

Molting. An exoskeleton must be shed periodically to allow for growth. Arthropods must cast off the old shell or skeleton, and form an oversized new one—a process called molting. Shedding the old shell occurs after secretion, but before hardening of the new one. Molting is accomplished by epithelial cells (surface tissue cells) secreting enzymes that weaken the old shell by eroding away the inner layers. Epithelial cells then secrete an outer layer for the new shell, which is resistant to the previously secreted enzymes. New, non-resistant inner layers are protected from the enzymes by the outer layer. When the thin, new shell is formed, the old, weakened shell splits open, the arthropod pulls itself out, and the old shell is called a cast. Immediately after molting, the new shell is soft, making arthropods vulnerable to predators. They usually remain hidden under rocks or burrowed in the bottom until the new shell has hardened. During the short period of hardening, the animal takes water or air into its internal spaces, allowing for a rapid increase in bulk. The arthropod eventually grows new tissues to fill the larger shell.

A six-year-old lobster (*Homarus americanus*) weighs about one-pound; it molts once a year, increasing its weight by 50% after the molt. Older lobsters molt about every two years. If a lobster, or another crustacean, has lost an appendage, it is gradually regenerated through successive molts.

Arthropods have a well developed nervous system and possess compound eyes. Most species are dioecious (separate sexes).

Class Merostomata (mer-o-sto′ma-ta) (Gr. *meros*, thigh, + *stoma*, mouth, + *ata*, characterized by). This ancient group is often referred to as "living fossils"; they have survived virtually unchanged for more than 360 million years. Only four species survive, three of them in Southeast Asia. *Limulus polyphemus*, an American species commonly known as the horseshoe crab, populates the Atlantic Coast from Maine to the Yucatan Peninsula on the Gulf of Mexico (Figure 1-22). It is not a "true crab" and is more closely related to spiders than crabs. True crabs (e.g., blue crab) have abdomens folded under the thorax, "like the tail of a frightened dog."

Class Crustacea (crus-ta'she-a)) (L. *crusta*, shell, + -*acea*, characterized by). The members of this class get their name from their hard exoskeleton. Most marine arthropods are crustaceans. They possess gills for respiration and two pairs of antennae. Thomas Pennant in 1777 in the fourth volume of his *British Zoology* separated the crustaceans from the insects with which Linnaeus had grouped them.

Subclass Cirripedia (sir-ri-pe'di-a) (L. *cirrus*, curl, + *pedis*, foot). Members of this class, such as barnacles, can curl their jointed appendages within protective calcareous plates (Figure 1-23, Color Plates 9c and 9d). Barnacles were classified as mollusks until John Vaughn Thompson, a British army surgeon and an amateur naturalist, described the development and striking metamorphosis of barnacles in 1830 in his *Zoological Researches and Illustrations*.

Figure 1-22. Horseshoe crab.

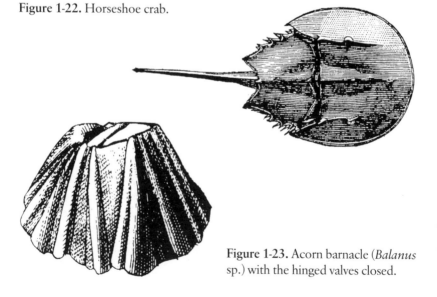

Figure 1-23. Acorn barnacle (*Balanus* sp.) with the hinged valves closed.

Subclass Malacostraca (mal-a-kos'tra-ka) (Gr. *malakos*, soft, + *ostrakon*, shell). Amphipods are small crustaceans, usually less than three-quarters-inch in length. The body is compressed from side to side with the head and tail curved downward. They do not have a carapace and have six pairs of abdominal appendages. The three hind pairs are longer and used for jumping; that behavior gives rise to common names like beach hopper and sand flea. The three front pairs are for swimming. Amphipods are found in decaying seaweed on the beach, on rocks, and swimming in tide pools. There are more than a thousand species and they provide a food source for many species of commercial and game fish (Figure 1-24, and Color Plate 10a).

Isopods are crustaceans, about the same size as amphipods, but with a dorsoventrally (top to bottom) flattened body (Figure 1-25). Like the amphipods, they do not have a carapace. The appendages are for walking and swimming. Isopods are important links in the food chains; many serve as scavengers.

Cumaceans are even smaller than amphipods and isopods, usually no more than a quarter-inch long, with a small carapace.

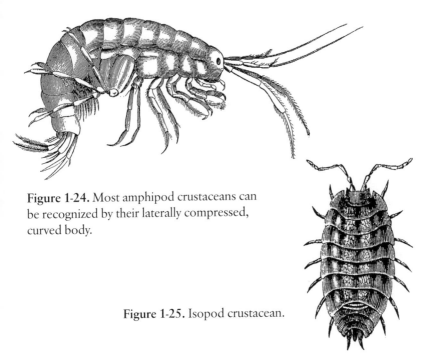

Figure 1-24. Most amphipod crustaceans can be recognized by their laterally compressed, curved body.

Figure 1-25. Isopod crustacean.

Decapods, the largest group of crustaceans, include the lobsters, shrimp, mole crabs, hermit crabs and the true crabs. The body is divided into two major parts, the cephalothorax (carapace) and the abdomen. They have five pairs of walking appendages; the first pair (chelipeds) is larger, and is usually provided with strong pincers filled with powerful muscles for defense and offense. Many decapods are of great commercial value (Figure 1-26).

Phylum Bryozoa (bry'o-zo'a) (Gr. *Bryos*, moss + *zoon*, animal)—also known as Phylum Ectoprocta (ek-to-prok'ta) (Gr. *ektos*, outside + *proktos*, anus). The phylum name bryozoa means moss animals, because the

Figure 1-26. A true crab (A) and a lobster (B) are examples of decapod crustaceans.

A

B

animals form encrusting or bushlike colonies on hard substrates (e.g., rocks, shells, pilings), seaweeds, and eel grass. The colonies consist of very small (less than ½₂ of an inch) interconnected individuals called zooids. Zooids are encased in external rectangular, round, or vase-shaped calcareous shell-like cases with an opening to extend feeding tentacles, which encircle the mouth. The anus is located outside the circle of tentacles, accounting for the phylum name ectoprocta (Figure 1-27).

Bryozoans grow in plantlike colonies and are not always easy to distinguish from hydrozoans. Both groups of animals were classified as zoophytes (animals resembling plants). In 1830, John Vaughn Thompson revealed that they are more highly organized and named them polyzoans, meaning "many animals." German zoologists independently arrived at the same conclusion and called them bryozoa, "moss animals."

Nudibranchs (sea slugs) prey on bryozoans, plucking them out of their cases with a tongue-like radula. The cases of some zooids have a

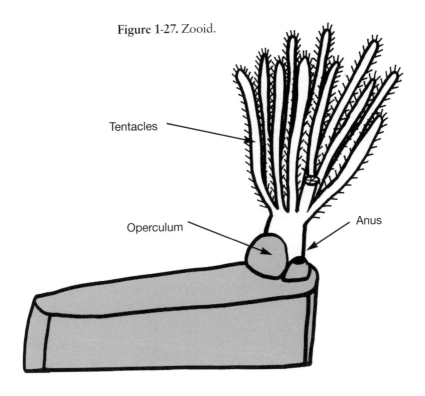

Figure 1-27. Zooid.

Tentacles

Operculum

Anus

jawlike operculum that snaps closed to discourage predators. Zooids do not have respiratory, circulatory, or excretory organs, probably because of the animal's small size. Neither do they have eyes or other specialized sense organs. Most bryozoans are hermaphroditic; released sperm are caught by the tentacles of another zooid and used to fertilize eggs.

Phylum Echinodermata (e-ki'no-der' ma'-ta) (Gr. *echinos*, hedgehog, + *derma*, skin). All echinoderms are marine; there are no freshwater species. They are relatively large animals when grown, the smallest being about one-half-inch in diameter.

Echinoderms and coelenterates (cnidarians) were initially classified as radiates, because of their radial symmetry. But symmetry is their only resemblence; echinoderms possess radial symmetry in the adult stage, although as larvae they are bilaterally symmetrical.

Most echinoderms have an undersurface called the oral surface, on which the mouth is centrally located; the upper surface is called the aboral surface (Figure 5-1). Internal calcareous skeletons composed of plates vary among the different classes. The skeletal plates (called ossicles) of sea urchins and sand dollars are fused, producing a rigid skeleton called a test. In sea stars and brittle stars, the skeletal plates are not fused, but articulate, forming a flexible skeleton and allowing movement of the arms. The skeletal plates of sea cucumbers have degenerated and the body feels soft. The name echinoderm means hedgehog skin; the bodies of hedgehogs are covered with spines as are most members of this phylum. The spines or tubercles of echinoderms project from the skeletal plates.

Echinoderms have a unique means of locomotion, the water vascular system. This is a water-filled system of internal tubes and external tube feet. By varying internal water pressure, the animal can extend or contract its tube feet, which usually end in small suction cups. The means of respiration (gas exchange) varies among the different classes of echinoderms. They do not have an excretory system, and their vascular system is so rudimentary as to be nonfunctional. Echinoderms are dioecious (separate sexes) but copulation does not occur; fertilization is external in the water.

Class Asteroidea (as'ter-oi'de-a) (Gr. *aster*, star, + *eidos*, form, + *-ea*, characterized by). Sea stars are star-shaped echinoderms with arms radiating from a central disk. The name sea star is preferable to starfish, because echinoderms are not related to fishes. In the study area, sea stars have five arms, but in other areas they can have as many as 44 and range in size from one-half-inch to more than three feet across. The shape and length of the spines that project from the surface vary among different species (Figure 1-28). The spines, covered with epidermis, have small scissorlike structures (called pedicellariae) at their base (Figure 1-29). It is believed that the cutting action of pedicellariae keep the sea star's body free of foreign material, such as seaweed, and possibly the removal of parasites. Also, near the base of the spines, small fingerlike processes (called dermal branchia), function as respiratory organs for gas exchange. An eyespot is located at the tip of each arm; it does not form an image but does distinguish light intensities.

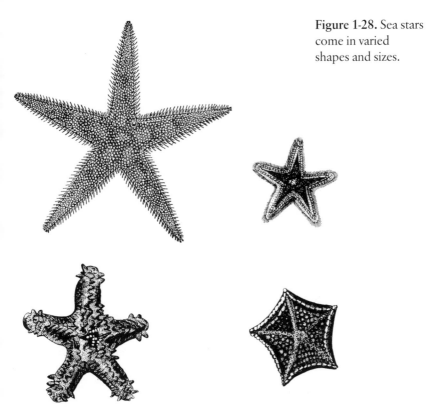

Figure 1-28. Sea stars come in varied shapes and sizes.

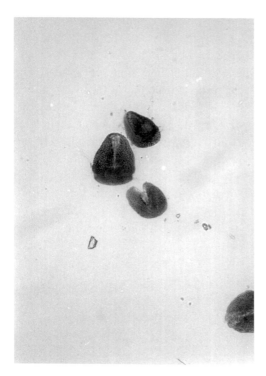

Figure 1-29. Sea star pedicellariae, magnified 40 times.

The sea star's water vascular system is used for both locomotion and food gathering. The description of the system presented here for sea stars is similar in other echinoderms. Water enters the water vascular system through the madreporite (a porous structure on the aboral surface that functions as a filter), passes through the stone canal into the ring canal, then the radial canals, and then the sac-like ampullae (Figure 1-30). Muscular contractions force the water from the ampullae into the tube feet, extending the tube feet (Figure 1-31). If the tip of a tube foot comes into contact with an object, the sucker will adhere to it. Muscular contractions in the wall of the tube foot force water back into the ampulla, shortening the foot. That action in many tube feet of one arm will slowly pull the sea star toward the object the suckers have attached to.

The water vascular system is also used to open the valves of mollusks such as oysters and clams. Instead of the alternating "push-pull" system used for locomotion, a slow steady contraction of the muscle tissue in the wall of the tube foot results in a constant pull, and the "tug

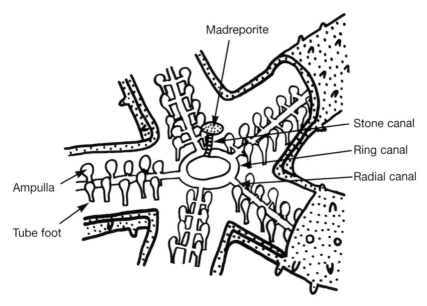

Figure 1-30. The water vascular system of canals and tube feet of a sea star.

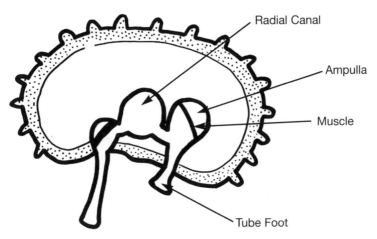

Figure 1-31. A cross-section of one arm of a sea star showing an extended and a retracted tube foot. Each ampulla has a muscle that contracts to force water into the tube foot, extending it.

of war" between predator and prey has begun. The bivalve's adductor muscles will exert a constant contraction, trying to keep the valves closed, but the steady contractions in the tube feet require less expenditure of energy and the sea star usually wins.

To see the tube feet of a sea star, carefully remove a sea star from the substrate, turn it over and observe the oral surface. Notice that each arm has a deep groove, the ambulacral groove, which extends from the mouth to the tip of the arm (Figure 5-1). The grooves contain the tube feet with the tips extending upward in an attempt to attach to something. If you remove a sea star from the glass wall of an aquarium, some tube feet will often tear off and remain attached to the glass. New tube feet will be regenerated.

A sea star's mouth leads to a stomach that consists of two parts, the cardiac region is everted through the mouth and over the food. Digestion actually begins outside the sea star's body and the partially digested food is then taken into the pyloric region of the stomach for further breakdown. The end products of digestion are passed into the digestive glands for absorption. Each arm has two digestive glands.

Sea stars are noted for their ability to regenerate new arms to replace lost ones. Their powers of regeneration are so great that a single arm, if a piece of the central disk is present, will generate a complete disk and new arms. Before this was known, oyster fishermen would take the sea stars dredged up with oysters, break them into pieces, and throw them back into the water. Instead of destroying their enemy they were increasing its numbers.

Class Ophiuroidea (o′fe-u-roi′de-a) (Gr. *ophis*, snake, + *oura*, tail, + *eidos*, form). Members of the class are called brittle stars or serpent stars. Like sea stars, they have arms radiating from a central disk. In contrast, however, the arms of brittle stars are longer and are separated from each other by an area of central disk, giving them a more vulnerable or brittle appearance (Figure 1-32). The snaky movements of the arms allow for locomotion; brittle stars are by far the most active echinoderms. Internal skeletal plates extend down each arm, giving a jointed appearance. The arms do not have ambulacral grooves. The madreporite plate is on the aboral surface of the central disk. The tube feet are reduced in number compared to a sea star and do not have suckers, but they are used to collect detritus for food. Also, the animals have jaws formed from large plates, and they feed on dead animals. Ten invaginations (slits) on the oral surface of the central disk contain tis-

Figure 1-32. Brittle stars.

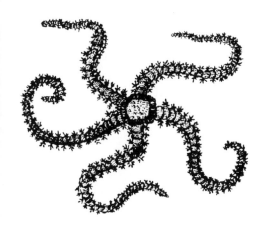

sue called bursae that function in respiration. Ciliated epithelium pulls water into and out of the small openings.

If a predator grabs one of the brittle star's arms, the echinoderm will cast it off and escape while the arm is being eaten. The missing arm will be regenerated as long as the central disk is intact.

Class Echinoidea (ek′i-noi′de-a) (Gr. *echinos*, hedgehog, + *eidos*, form). Members of the class are circular- or oval-shaped echinoderms with numerous movable spines and no arms. Large ossicles are fused together and form a test (a shell-like skeleton). Openings in the test allow the tube feet to protrude. The madreporite plate is on the aboral surface and the tube feet have suckers.

SEA URCHINS

Sea urchins are covered with long spines that aid their tube feet in locomotion (Figure 1-33). The tube feet are scattered all over the body between the spines and extend beyond the spines. The tests (does not include spines) of most species are about three inches in diameter, but a few Indo-Pacific species may grow to almost one foot in diameter. Sea urchins possess pedicellariae and respiration occurs by peristomal gills. When the animals die, the spines fall off and the tests frequently wash ashore (Figure 1-34). If the test is intact, the teeth can easily be seen on the oral surface. Gulls, crabs, fish, and other animals prey on sea urchins.

SAND DOLLARS

Sand dollars are close relatives of the sea urchins but are flattened, and the spines are much smaller (Color Plate 13c). Like the sea urchin, both the tube feet and spines make movement possible in sand, where they feed on microscopic organisms. Sand dollars do not have pedicellariae or peristomal gills, modified tube feet act as gas exchange structures. Bottom feeding fish, such as flounder, prey on sand dollars.

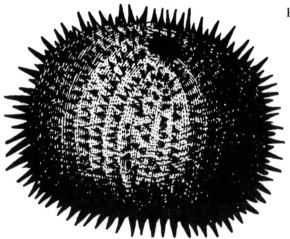

Figure 1-33. Sea urchin.

Figure 1-34. Sea urchins usually lose their spines as they wash ashore. The test (skeleton) reveals the close relationship to sea stars. The star pattern of openings, which allow tube feet to protrude, are reminiscent of the five arms of a sea star.

Class Holothuroidea (hol'o-thu-roi'de-a) (Gr. *holothourion*, sea cucumber, + *eidos*, form). Sea cucumbers are cucumber-shaped echinoderms with no spines or arms. Microscopic ossicles are buried in the thick, muscular wall; the madreporite plate is within the body. The tube feet possess suckers. Gas exchange is accomplished by a system of internal tubules called respiratory trees. A sea cucumber has two respiratory trees, and water enters and leaves the tubules through the anus. Some species possess additional tubules (called tubules of Cuvier) attached to the base of the respiratory trees. If the animal is attacked, the anus can be directed toward the predator, and the tubules of Cuvier are shot out through the anus. The tubules are released from their attachment and will be regenerated in about six weeks. The tubules are coated with a sticky substance and can entangle the crab or other predator, allowing the slow-moving sea cucumber to escape. Also, the predator, if not tied up by the tubules, may stop to eat them, allowing the echinoderm to escape.

Sea cucumbers have a feeding device that is unique among echinoderms. A cluster of 10 to 30 tube feet around the mouth are modified to form large tentacles (Figure 1-35). The tentacles brush over the bottom and pull detritus into the mouth.

When a sea cucumber is disturbed, it will contract its tube feet and feeding tentacles, making it difficult to recognize that the animal is an echinoderm.

In the Orient, dried sea cucumber (trepang) is considered a delicacy.

Figure 1-35. Sea cucumber; note the feeding tentacles on the left.

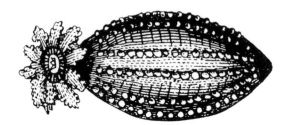

Phylum Chordata (kor-da′ta) (L. *chorda*, cord, + *ata*, characterized by). This phylum has received, by far, the most study from zoologists because it includes fishes, reptiles, amphibians, birds, and mammals. Chordates exhibit bilateral symmetry, without obvious segmentation. They possess a dorsal hollow nerve cord and a stiffening rod, the notocord (from which the phylum derives its name) at some time during their life cycle. Gill slits, opening from the pharynx, are also present at some time in the life cycle. Members of this phylum range from the often overlooked tunicates to the largest marine inhabitants, whales. The blue whale is the largest animal ever to inhabit the earth.

Subphylum Urochordata (u′ro-kor-da′ta) (Gr. *oura*, tail + L. *chorda*, cord, + *ata*, characterized by). Members of this subphylum are commonly known as tunicates, with little resemblance to other chordates. It is hard to believe they are such advanced, complex animals. Most adults are somewhat barrel-shaped, often form colonies, and are usually found attached to a hard substrate. The microscopic, free-swimming larval stage looks like a tadpole and has a notocord and nerve cord. These two chordate characteristics are lost in the sessile adult stage, while the pharyngeal gill slits are only found in the adult (Figure 1-36).

The tunicate's body is surrounded by a special sac-like mantle (the tunic) in most species, giving rise to the name tunicate. The tunic provides support and protection for the soft body parts; it varies in thickness from a soft delicate consistency to one that is tough, similar to cartilage. The constituents of the tunic can include cellulose (called tunicine), proteins, inorganic compounds, and even calcium salts in the forms of distinct spicules.

Figure 1-36. A diagramatic representation of a tunicate larva (A) and an adult (B). Note the notocord and nerve cord in the larva, both of which are reduced or absent in the adult.

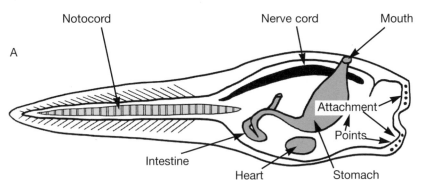

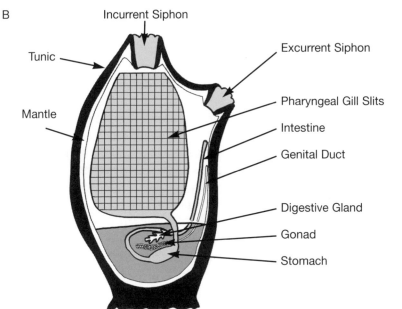

Tunicates are filter feeders and the adults have two siphons. Cilia on the pharynx beat in such a manner that water is pulled in the incurrent siphon, passing over the mucus-covered pharynx, through the gill clefts, and out the excurrent siphon. The mucus with the trapped food is passed to the digestive system. Interior tentacles near the incurrent siphon inhibit large, destructive objects from entering with the water.

Some species are solitary while others produce colonies. The individual animals (zooids) in the colonial forms are interconnected by a threadlike structure called a stolon (Color Plate 15a). The zooids of some colonial species have an incurrent siphon but share an excurrent siphon with neighboring animals.

Tunicates are hermaphroditic (male and female reproductive organs are present in the same animal), but self-fertilization does not occur. Gametes (sex cells) are released into the water and the sperm of one animal fertilizes the eggs of other tunicates.

Subphylum Vertebrata (ver'te-bra'ta) (L. *vertere*, to turn + *ata*, characterized by). A notocord is present only in the embryo and is replaced by a chain of vertebrae. The central nervous system, particularly the brain, is very highly developed. The cranium (a skeletal case) surrounds and protects the brain.

FISHES

Few descriptions of fish species are presented in this guide because beachcombers see few live specimens. But they are so important to the marine environment that the classes will be described briefly.

Fishes are the undisputed masters of their environment; no other group threatens their domination of the seas. They are perfectly adapted for aquatic existence. Most have a swim bladder to regulate buoyancy. Their fins function well as brakes and rudders. Fishes possess excellent olfactory and visual senses and a unique system in their lateral lines, which is sensitive to water currents and vibrations. Their gills have evolved into the most effective system in aquatic animals for gaseous exchange. Many species have developed complex behavorial patterns for reproduction with courtship, nest building, and care for the young. These are only a few of the adaptations that have contributed to the mastery of their aquatic environment.

---

*A word about the sometimes confusing usage of "fish" and "fishes." "Fish" is plural when it refers to more than one fish of the same species; "fishes" refers to two or more species.

Class Chondrichthyes (kon-drik'thee-eez) (Gr. *chondro*, cartilage + *icthys*, fish). Sharks, skates, and rays are representatives of this class. They possess a cartilaginous skeleton, with five to seven pairs of separate gills. The scales are small and do not overlap (Figures 1-37 and 1-38).

Class Osteichthyes (os'te-ik'thee-eez) (Gr. *osteon*, bone + *ichthys*, fish). The class includes more species than any other class of vertebrate. Members of the class range from the familiar flounder to the unusual sea horse. The skeleton is composed of cartilage and bone. The gills are reduced in number and are covered with a thin bony structure, the operculum. Scales are larger and overlap one another (Figures 1-39, 1-40, and 1-41).

Figure 1-37. Sharks.

Figure 1-38. Sting ray.

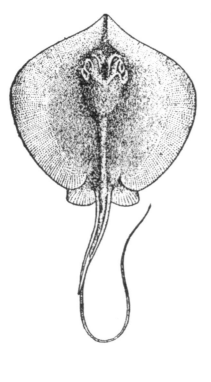

Figure 1-39. Sea horse.

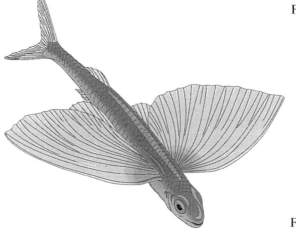

Figure 1-40. Flying fish.

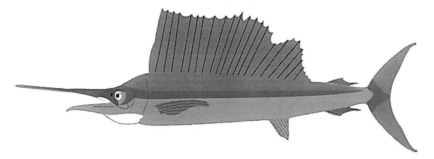

Figure 1-41. Sailfish.

REPTILES

Class Reptilia (rep-til'e-a) (L. *repere*, to creep). The class of chordates that are usually called reptiles. They are distinguished by scales on the skin. The only marine reptiles found in the study area are the sea turtles. Snakes and lizards are occasionally seen in the sand dunes. Being cold-blooded, reptiles can leave their eggs to hatch without incubation.

BIRDS

Class Aves (ay'veez) (L. *avis*, bird). The class of chordates that are usually called birds. They are distinguished by the presence of feathers on the skin. Birds are warm-blooded (homeothermic), but fossil evidence indicates that they may have evolved from reptiles. They have several characteristics in common with reptiles. Birds possess scales on their legs and sometimes around the base of the bill (Figure 1-42). Feathers are thought to be modified scales; wings are thought to be modified anterior legs.

Feathers define birds; they are the only animals that possess those "modified scales." Feathers, like the fur of mammals, provide excellent insulation. Their color patterns serve as camouflage, especially for the young, to protect against predators. As warm-blooded animals, birds must incubate eggs for them to hatch.

To master flight, certain body modifications were necessary. Avian bones are thin and more hollow to minimize weight, the breastbone (sternum) is enlarged and functions as an anchor for the powerful breast muscles that are used for flight.

Most birds have an oil gland near the tail; when they preen their feathers they add the secreted oil to the feathers, waterproofing them. Most birds molt (shed their feathers) once a year.

Birds have bills (two mandibles or jaws), also called beaks, but no teeth.

Seashore birds have salt glands near the nares (nostrils) in their bills. After a bird drinks salt water the gland accumulates salt and a salt solution drips out of the nares. By carefully observing a seashore bird such as a gull, you may see a droplet of the secretion hanging on the tip of its bill.

Figure 1-42. Sandpiper.

MARINE PLANTS

Seaweeds

Marine algae (pronounced al-jee), commonly known as "seaweeds," are not really weeds. Weeds have vascular tissue (specialized tissue for transporting products throughout the plant); they are flowering plants with true roots, stems, and leaves. Algae (singular alga) do not have vascular tissue, and their plant body (thallus) lacks differentiation into roots, stems, leaves, and flowers. Algae generally have three main parts—a blade or frond, a stem-like stipe, and the holdfast that secures the plant to its substrate. The holdfast is either rootlike in appearance, or disc-like expansions of their tissues. The holdfast is quite effective. Only the heavy seas of a violent storm can tear the plants loose.

Roots are not necessary. Instead of extracting minerals from the soil as a land plant must, algae are bathed almost continually by the sea. It provides all the needed minerals. A rigid supporting stem is also unnecessary. The stipe, when present, is flexible, allowing the plant to yield to waves. As a result, the alga's structure is simple, merely branching filaments or blades rising from a stipe or a single holdfast.

Algae have developed five major body types: unicelluar, colonial, filamentous, membranous, and tubular. The simplest is unicellular (single cell). A slightly more complex level of organization is the colonial type, an association of cells of the same species. They may be undifferentiated or contain several kinds of cells illustrating differentiation, specialization, and division of labor. Cell division occurring in one direction results in chains of cells called filaments. When some cells divide in a new direction, branches are produced.

Cell division in two or more planes results in a membranous structure, one or several layers of cells thick. In some algae, the plant body forms a tube.

To identify some algae, it is necessary to cut sections of the thallus to observe it with a microscope. This can be done with a single-edge razor blade. The specimen can be held with the index finger on a glass slide and sliced with the blade against the finger nail. Fairly thin sections can be produced in this manner with a little practice.

The major groups (divisions) of algae are segregated on the basis of pigmentation, storage products, chemical nature of the cell wall, and flagellar (long, hairlike projections from the cell) number and insertion. It is the pigment (color) differences that are initially the most evident.

A major factor in their locations on the substrate is the amount of light to which they are exposed, some growing at greater depths than others.

The layman's use of the term seaweed often implies that it is a useless, or even noxious plant of the sea. It often litters a lovely beach, fouls the bottom of a boat, or becomes entangled in fishing line. However, marine algae are beautiful additions to the marine environment. They are also beneficial, both to marine animals and humans. They provide oxygen (a waste product of photosynthesis), food, and shelter for a host of marine organisms. They are responsible for an estimated 60% of the oxygen present in the air we breathe.

About a dozen different types of seaweed are harvested commercially. In the Orient, especially Japan, they are used extensively in the diet. In the Western World, little use is made of seaweed for human food. We are not adventurous when it comes to eating seaweed. However, Irish moss (*Chondrus crispus*) has been used since Colonial times as a thickener for soups and desserts. Europeans use considerable amounts as animal feed and fertilizer. Cattle and sheep often graze on seaweed at low tide in Scotland and other European countries. In the United States, only small amounts of seaweed meal are produced for animals. Some areas of New England use seaweed as fertilizer.

Products derived from seaweed include iodine, sodium chloride, and potassium chloride, and are produced by charring kelp (large seaweeds). Fermentation produces acetone, ethyl acetate, and other solvents. Other products derived from marine algae are the jellylike phycocolloids. Agar is produced from red algae and algin from the kelps. Agar is used extensively for the culturing of bacteria and fungi in the laboratory. Algin has many uses, including waterproofing cloth and improving the texture of ice cream, chocolate milk, and other dairy products.

Harvesting techniques vary, usually dictated by depth. In the intertidal zone, seaweed can be gathered by hand or raked. An experienced harvester can gather as much as 1,000 pounds on a tide by such manual methods. Scuba diving has become commonplace in marine botanical investigations.

The marine algae serve man in many ways, but it is far more important that they are indispensable sources of food, and places to hide from predators for a host of marine animals.

Collecting Marine Algae

The collection of algae should be planned to coincide with the hours of a falling tide. Plan to arrive at the collection site before low tide, and work out with the ebbing water. Collecting may be easily accomplished by walking along the beach, gathering up the specimens that have broken loose and washed ashore, especially after storms. Some plants may have been on the beach for a long time and be bleached out by the sun. Their loss of color can make identification more difficult. By wading in shallow bay waters, wearing shoes for protection, many detached floating specimens can be collected. Collectors should visit a rocky shore or jetty at low tide for algae that are still attached if the weather, particularly wind, permits. The algae can be placed in separate small Ziploc plastic bags with a little seawater to keep them moist. The bags should be carried in a plastic bucket. Glass containers should not be used, to avoid injury in case of a slip or fall. Rocks, often slippery due to the growth of algae, can be dangerous. Corrugated rubber-soled shoes should be worn while collecting on rocks.

The collected specimens may be examined by placing them in a shallow white enamel pan containing seawater. Algae that grow upon another plant are called epiphytes. Large algae should be carefully inspected for epiphytes; some may be microscopic. If the plants are to be preserved, 70% ethyl alcohol or rubbing alcohol is satisfactory. The bottles containing the preserved specimens should be kept in the dark to retard loss of color.

If the algae are to be pressed in a plant press, float the specimen in a pan of seawater. Place the herbarium paper (5 × 8-inch index cards will do for small specimens) beneath the plant and spread the alga out with a teasing needle so that it exhibits its natural features. With one hand under the paper, gently raise it out of the pan, allowing the water to drain off. With the other hand, continuously spread the alga as the water flows from the paper. Less delicate forms can be placed directly on the paper without floating, but should still be arranged.

Place the mounted specimens in a plant press (two ½-inch plywood boards with rocks or bricks for weight can be used instead of a commercial press), on a piece of felt paper if it is available; if not, use several pieces of newspaper. Cover the alga with a piece of smooth cotton cloth or wax paper to prevent it from sticking to the two or three sheets of newspaper that is then placed on top of the covered specimen. Be careful when removing the cloth, sometimes it will stick to the specimen. Place corrugated cardboard between each preparation to allow air to circulate through the press. The correct layering sequence should be cardboard, felt paper or newspaper, mounted specimen, cloth, and newspaper.

When the press is loaded and is ready to be strapped or screwed down, only moderate pressure should be applied. Too much pressure will result in corrugated impressions on the specimens. Room temperature is satisfactory, but the newspaper should be changed daily for the first week, then every third or fourth day until the specimens are completely dry. If the specimens are removed too soon the paper will curl as it finishes drying. Many specimens do not adhere well to paper; they will have to be glued to the sheet after drying.

All collected specimens should be labeled with the scientific name of the alga, the date of collection, the habitat, the locality, the name of the collector, and the person who makes the identification of the specimens.

Division Chlorophyta (klor'o-fi'ta) (Gr. *khloros,* green + Gr. *phyton,* plant). The green algae are the most diversified of all the algae. All body forms (unicellular, colonial, filamentous, membranous, and tubular) are present. These algae are green in color because the green pigment chlorophyll is the predominant pigment (Color Plate 3a). Green algae are generally found growing attached to objects that are exposed at low tide. The number of species large enough to be conspicuous is small compared to the brown and red algae. The shape and number of chloroplasts (intracellular structures that contain chlorophyll) are useful in identification. Conspicuous in the large chloroplasts are dark structures called pyrenoids which are associated with starch formation.

Division Phaeophyta (fa'i-o-fi'ta) (Gr. *phaios,* dusky + Gr. *phyton,* plant). The brown algae have a characteristic brownish color because the brown pigment fucoxanthin masks the green pigment chlorophyll. On occasion, the amount of brown pigment is not enough to mask the chlorophyll, and the plant appears green. This can confuse an inexperienced collector. Most species of brown algae are found in the mid and lower intertidal zone and below the low water mark. The large members of this division that develop parts resembling stems and leaves are called "kelps" (Figure 1-43).

Division Rhodophyta (ro'do-fi'ta) (Gr. *rhodon,* rose + Gr. *phyton,* plant). The characteristic red color of the group is due to the presence of the pigments phycoerythrin (red) and phycocyanin (blue), which mask the green chlorophylls (Color Plates 2d and 6c). Like the brown algae, color is sometimes an unreliable criterion because some reds have a green or brown color. The mixing of red and blue pigments often produces a purple color. Most red algae prefer not to be exposed to the air; they are usually found below the low water mark. They are the most common of marine algae in number of species.

Their size and complexity varies, but most red algae are small, compared to the kelps. Some species are beautiful when mounted on white paper as herbarium specimens.

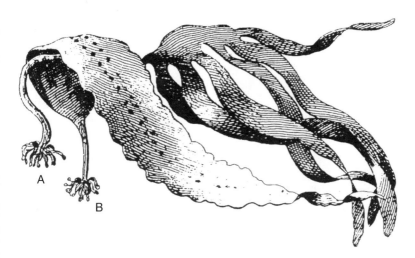

Figure 1-43. Two kelps, *Laminaria digitata* (A), with the divided blade, and *Laminaria saccharina* (B), the blade is long, undivided and has distinctive undulations.

Flowering Plants

Division Anthophyta (an'tho-fi'ta) (Gr. *anthos*, flower + *phyton*, plant). Flowering plants are represented in the marine environment by the sea grasses, marsh grasses, beach grasses, and sand dune plants.

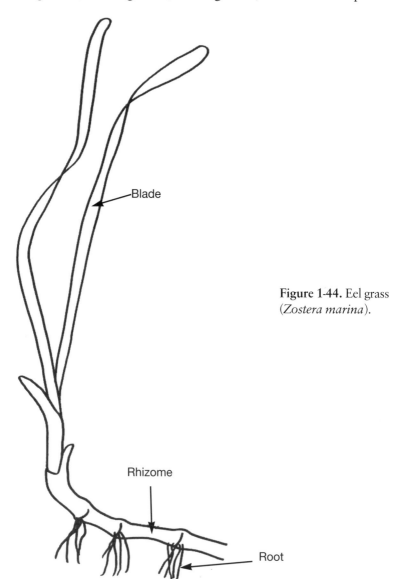

Blade

Rhizome

Root

Figure 1-44. Eel grass (*Zostera marina*).

They are not algae; they possess vascular tissue and true roots, stems, and leaves. Vascular tissue consists of tubelike vessels that conduct fluids up and down the plant.

The sea grasses consist of a branched rhizome (underground, horizontal stem) that produces erect shoots on the upper surface and roots on the undersurface (Figure 1-44). Each shoot consists of a very short stalk that bears long, green, bladelike leaves. The small flowers of sea grasses are hidden in leaf sheaths. Pollen, which contains the male sex cell, is carried by water currents. Marsh grasses have large conspicuous flowers that can be used to distinguish the various species. Sea grasses grow wholly submerged while most marsh grasses are wholly or partially submerged only at high tide. Beach grasses and sand dune plants are seldom submerged.

Terrestrial flowering plants are thought to have evolved from marine ancestors, but the sea grasses have migrated from the land back to the sea.

Vast areas of sandy and muddy bottom, in bays and estuaries, are covered with eel grass (*Zostera marina*), the only sea grass in the study area covered by this guide. Eel grass is important because the rhizome stabilizes the substrate and protects against erosion, and the plants are a food source for birds and many marine animals. Not only does the dense foliage provide protection for many immature marine animals, but the decomposing leaves are a major source of detritus, which nourishes planktonic organisms far out at sea. Fish, shellfish, crabs, and many other marine animals find shelter in beds of eel grass. Many algae are epiphytic on eel grass, and densely cover them, providing food for many fish, crabs, shrimp, and other animals.

2

Beach Basics—Features and Creatures

Ever drifting, drifting, drifting
On the shifting
Currents of the restless main
Till in sheltered coves, and reaches
Of sandy beaches,
All have found repose again.

—Henry Wadsworth Longfellow

INTERTIDAL ZONE

The coastal strip that separates land from sea is called the intertidal, or littoral zone. It is, by far, the most accessible and productive place for a beachcomber to observe marine life. The area is defined by the highest of high and the lowest of low tide levels. It contains a broad spectrum of habitats from sand beaches, rocks, and marshes to man-made docks, jetties, and breakwaters. The physical conditions, the most variable in the entire marine environment, compel its inhabitants to adapt to daily exposure to air and changes in temperature as the tide rises and falls.

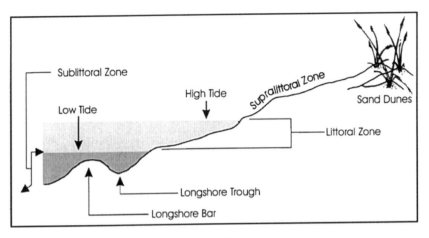

Figure 2-1.

Below the intertidal zone is the sublittoral zone, which extends, with gradually increasing depth, to the edge of the continental shelf. The supralittoral zone is above the high tide mark, technically dry land, but is often wet by the splash and spray of breaking waves (Figure 2-1).

A leisurely stroll is all that is needed to comb the productive intertidal zone. It is exposed twice a day for variable lengths of time. In fact, the tide is a wave moving through the ocean, with high tide the crest and low tide the trough. As the tide rises and falls it reaches different points along the shore at different times. Beachcombers need to know about the weather and the tide to make effective use of their time. A local tide table, available in bait stores and newspapers, will help to determine when low tide occurs; it provides the maximum area for observation. An annual tide table, for the entire Atlantic Coast, can be obtained from the Superintendent of Documents at Washington, D.C.

Tides

For centuries before the time of Christ it was known that the moon was somehow related to tides. Arabian philosophers of the sixth century A.D. attributed the flooding tide to the moon's heat, warming the ocean and causing the water to expand. However, the true nature of the cause-and-effect relationship between moon and tides was not understood until 1867, when Sir Isaac Newton published his *Philosophiae naturalis prin-*

cipia mathematica, which stated his theory of universal gravitation. Newton, who also first described wave motion mathematically, explained that the tides were the result of the gravitational pull of the sun and moon on the earth's water, producing bulges. The moon is the more powerful force because it is so much closer than the sun, which is 390 times farther away. The moon orbits around the earth, and both orbit about the sun. The gravitational attraction of the moon and sun is counterbalanced by centrifugal forces. The earth, in moving around the sun, is held in orbit by a balance of two forces—gravitational and centrifugal. Gravity pulls the earth and sun together, and centrifugal hurls them apart.

Both forces produce bulges of the water mass on opposite sides of the earth (Figure 2-2). The moon rotates once around the earth every 24 hours and 50 minutes (a lunar day). Thus, a location in the study area moves through each of the ocean bulges, so there are two daily high tides and two daily low tides (actually over a 24-hour and 50-minute period). Each low tide or high tide follows the previous low or high by about 12 hours and 25 minutes.

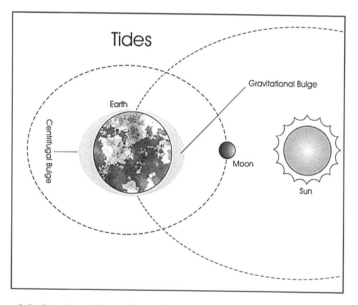

Figure 2-2. Gravity and centrifugal forces produce bulges of the water mass on opposite sides of the earth.

A "spring tide" occurs about every two weeks, with a full moon or a new moon producing the maximum low tide for the largest study area (Figure 2-3). The term "spring tide" is not related to the season, but is derived from the Saxon "sprungen," referring to the brimming fullness of the water causing it to spring up. During both full and new moon phases, the sun, earth, and moon are in line, mutually reinforcing the gravitational attraction rather than behaving independently. During full moon, the earth lies directly between the sun and moon; at new moon, the moon lies between the earth and sun. At those times, and for a few days before and afterwards, tides are higher and lower than at other times. The least amount of water movement, a "neap tide," is produced when the sun and moon are at right angles with the earth, also about every two weeks. Then, with both the sun and moon attracting the water, interfering with each other, the crest is diminished (Figure 2-3). The word "neap" has a Scandinavian root meaning "barely touching"

Figure 2-3. The moon rotates about the earth every 29.5 days; it is aligned with the sun twice a month (A) and is at right angles twice a month (B).

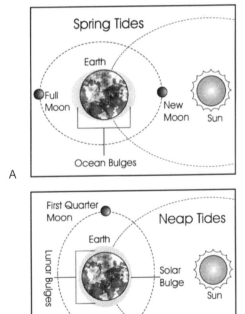

or "hardly enough." That entire lunar cycle requires 29.5 days. On the Atlantic Coast, the greatest tides occur during spring and autumn equinoxes.

Slack water is a term used for the period of an hour or so following high or low tide, when the water is neither falling nor rising. Slack water at low tide, before the turn of the tide, is the best time for exploring the low-level area of a sand beach or rock jetty. Slowly work your way up the beach or rocks as the water moves in.

Wind-Produced Waves

The commonly seen waves on the surface are caused principally by wind. However, submarine earthquakes and tides also cause waves.

The sun's heat generates air currents that form winds which push against the water's surface. The absorption of wind energy by the water produces waves (Figure 2-4). As wind speed increases, larger waves form. Their size depends on three factors: strength of the wind, length of time it blows, and fetch (extent of water over which a wind can blow). For example, the waves on the north shore of Long Island, formed from wind traveling over Long Island Sound, are smaller than those on the south shore, where the wind has the vast expanse of the Atlantic to travel over.

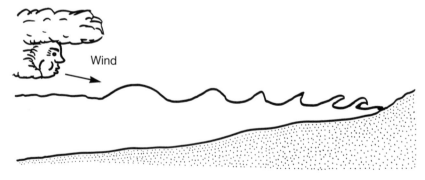

Figure 2-4. Wind energy produces waves.

A wave is energy transfer *through the water*, waves *do not* carry along the water, except when they are close to shore. As a wave travels, water molecules move in a circular motion, as evidenced by a gull floating offshore; it bobs up and down as waves pass under it, but does not move horizontally (Figure 2-5).

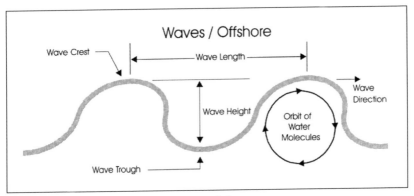

Figure 2-5.

When a wave nears the beach it begins to "feel" the bottom; friction decreases its speed, the wave increases in height, and the back of the wave crowds the front, piling the water higher until it breaks, forming surf. The slope of the bottom offshore is the primary factor in determining whether a breaking wave is a "spiller" or "plunger." Gently sloping bottoms usually produce breakers, with the crest gradually spilling over the top for some time. The greater the bottom slope, the more abruptly a wave will crest and plunge into the surf. Spillers give a longer, but less impressive ride, to surfers.

The sea foam is composed of air bubbles separated from each other by a film of liquid. Most bubbles are caused by wind waves, but rain may also produce them (Figure 2-6). When bubbles burst, salt spray is released into the air. It is believed that most of the airborne salt come from bursting bubbles.

Sand beaches are dynamic, not static. Changes are expected. Sand beaches are the most temporary and variable of all the oceanic regions. From day to day small changes may be noticed, but from season to sea-

Figure 2-6. Sea foam produced by breaking waves.

son the changes are quite pronounced. Storm-produced waves generate a tremendous force and every year account for considerable beach erosion. The beaches in the study area are predominantly unsolidated sand and gravel. Storms such as "nor'easters" produce up to 20-foot-high waves combined with a 3-foot rise in sea level due to the low pressure. Such high-energy conditions will cause severe erosion of the unconsolidated sand and gravel coast (Figure 2-7).

Waves generally approach a shoreline obliquely rather than head-on. The oblique approach shifts sand along the beach, and produces a long-shore current that runs parallel to the shoreline, away from the approaching waves (Figure 2-9). The sand that is transported in the current is referred to as littoral drifting. Structures constructed along the shore, such as rock jetties to protect inlets and groins to protect against erosion, will trap sediment on the upcurrent side, but erosion often becomes a problem downcurrent.

Occasionally, the water in a longshore current will flow back to the sea through a narrow opening in an offshore sand bar, forming a "rip current" (Figure 2-10). Swimmers often refer to it as an undertow, but

Figure 2-7. Montauk Lighthouse, at Montauk Point, the easternmost point of Long Island (N.Y.), is the island's most photographed landmark. When the lighthouse was constucted in 1795, it was 297 feet behind the bluff; now it is as little as 50 feet in some spots.

Figure 2-8. Lighthouses are unique in design or paint pattern so that mariners would be able to distinguish between them and know their position along the coast. Cape Hatteras Lighthouse on Hatteras Island (left) has a spiral black stripe, while Bodie Island Lighthouse (right) just a few miles up the coast has two circular black stripes. Built in 1823, Ocracoke Lighthouse (center), a few miles to the south of Cape Hatteras Lighthouse, is the oldest operating lighthouse in North Carolina. Cape Hatteras Lighthouse, built in 1870, is, at 208 feet, the tallest in the United States. Bodie Island Lighthouse was built in 1872. (Courtesy of the National Park Service.)

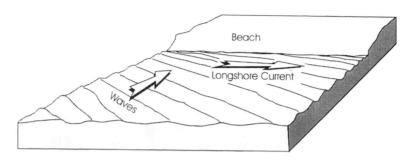

Figure 2-9. Waves in the surf zone produce a longshore current that transports sand down the beach.

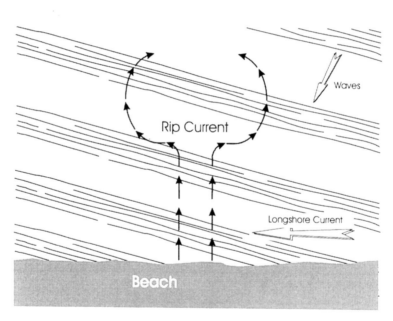

Figure 2-10.

the name is misleading; there is no downward movement of water, only offshore flow. Swimmers caught in a rip current often panic because water movement is usually severe and a strong swimmer attempting to swim against it may become exhausted and drown. If caught in a rip current, relax and let the water carry you out until its speed diminishes, then swim parallel to shore and return at a different point. The other option is to swim through (again parallel to the beach) the narrow rip current as you are being pulled offshore.

SAND BEACHES

Origin

The most familiar intertidal zone is the sandy beach. Sand beaches dominate the Eastern Seaboard between Cape Cod and Cape Hatteras. They consist mainly of quartz and feldspar, which result from the weathering of granite, then are carried to the shore by rivers or surface runoff. The major source of sand for the region from Long Island to Cape Cod is the deposit left by a glacier retreating at the end of the Ice Age. At one time, a sheet of ice several hundred feet thick blanketed Connecticut and Long Island Sound. It reached its southernmost limit, and held that "standstill" position for thousands of years. As the temperature warmed, the ice melted and the glacier receded, leaving behind clay, silt, sand, gravel, rocks, and large boulders that were incorporated in it. A mound of such debris is called a terminal moraine.

The first moraine formed the south fork of Long Island, Block Island, and most of Martha's Vineyard. As the glacier retreated, a second "standstill" of several thousand years occurred, followed by a resumption of the glacier's retreat to the north. That produced the north fork of Long Island, Fishers Island, and part of Cape Cod (Figure 2-11).

Changes

Because much of the earth's water was contained in the ice sheets, sea level was about 300 feet lower than it is today. The melting of the glaciers at the close of the Pleistocene period (about 10,000 years ago) increased the volume of water in the ocean. Sea level rose and river valleys that once flowed into the Atlantic, a considerable distance seaward of the present coastline, were "drowned," producing Narragansett Bay, New York Harbor, Delaware Bay, and Chesapeake Bay. About 4,000 years ago, sea level had reached the approximate position it is today.

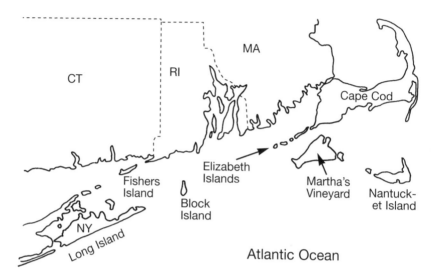

Figure 2-11. Cape Cod was formed when three lobes of the ice sheet converged. Long Island is the result of two moraines, producing a north and south fork at the eastern end.

Winds, currents, and waves have reworked and redistributed the glacial deposit in a process that even now continues.

Some beaches are formed from sands deposited offshore on the continental shelf and carried landward. Another source of beach materials is the erosion of seashore cliffs and headlands by wave action, such as occurs on the north shore of Long Island, imperiling the homes of local residents. The finer soil particles are carried away, leaving the larger rocks scattered on the beach. When a beachfront resident sees his backyard wash away, or when a storm tide breaches a barrier island and creates another inlet to the bay, it is the natural "shaping" process still at work.

BARRIER BEACHES

That shaping process produced barrier beaches. Barrier beaches include spits and barrier islands. Barrier islands are linear islands that parallel the mainland. Barrier beaches are dynamic, not static, geologic features, built by the action of waves and currents. Wind-driven ocean waves move offshore sand landward, first producing a sand bar. This continuous action and the longshore current, carrying sand parallel to the mainland, continues the buildup of the sand bar until it is above sea level. The exposed sand, the beginning of a barrier island, is separated from the mainland by a shallow bay, or lagoon. Once above the reach of spring tides, wind builds the island higher and wider, and eventually forms dunes. The establishment of plants stabilizes the dunes, creating a barrier island that bears the brunt of waves and tides, and provides storm protection to the mainland.

Barrier islands migrate toward the mainland as wind and waves continuously move sand in that direction. As it builds up, the wind blows the sand into the bay between the barrier island and the mainland. That buildup of sediment contributes to the formation of marshland that encroaches on the bay. As vegetation develops in the marsh, it slows the movement of water and traps sediment, filling in more of the bay. As the island moves toward the mainland, the marshes will be covered. Over the very long term, the barrier island will then be part of the mainland (Color Plate 1a).

Sand spits, off the mainland or by inlets, form in a similar manner, and are fed by wave action and the longshore current. On the south shore of Long Island, sand spits develop southwestward, the direction of the littoral drift of sand. The barrier islands of Long Island's eastern end are believed to have initially formed as a sand spit attached to a glacially developed lobe at Southampton. The spit breached in 1931 to form Fire Island and again in 1938 to form Westhampton Barrier Island.

It is estimated that there are 295 major barrier islands in the 18 coastal states from Maine to Texas.

SAND DUNES

A sand dune is a ridge of wind-deposited sand. The windward side usually has a gentle slope, while the lee side is usually steeper. On a day when the wind is blowing hard from offshore, look down the beach and, in the distance, the air looks hazy from wind-driven sand (Figure 2-12). The first dune inland from the beach is the primary dune or foredune.

Beyond the foredune is the dune-and-swale community. The plants that live in this harsh desertlike environment are subjected to salt spray on the windward side of the dunes, and possible submergence during severe storms should storm waves breach the dune. Plants must be able to survive the sea, salt, nitrogen poor soil, erosion, and desiccating heat.

The sand dune environment is dominated by vascular plants. One of the first plants encountered, near the base of the foredune, is dusty miller (*Artemisia* species) (Figure 2-13). The common name is derived from its appearance of looking as though it had been dusted with white powder. A dense cover of white, hairlike fibers that insulate the plant from heat is responsible for the dusty white appearance. The plant has silvery foliage that is soft to the touch and tiny yellow flowers in June, and grows to about two feet high (Color Plate 1d). Another common name is beach wormwood. Cultivated as a garden plant, it was imported from Europe, found its way to the beach, and has spread from New Jersey to Nova Scotia.

Beach grass (*Ammophila breviligulata*), also known as dune grass, is usually the dominant plant in the dune area nearest the beach. The sci-

Figure 2-12. A wind-produced "sand mushroom" on a North Carolina beach. (Photo courtesy of the National Park Service.)

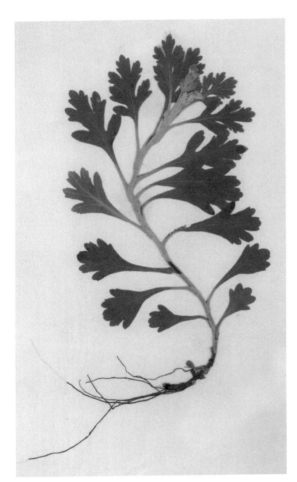

Figure 2-13. Dusty miller (*Artemisia* sp.). The specimen is 8 inches across.

entific name means "sand-lover" and the plant thrives in pure sand. It adds beauty to the dunes and grows in stands or "fountains." Beach grass has long, narrow, fibrous blades, with drooping tips that swing in the wind forming circular patterns in the sand. Its compass-like circles give rise to another common name, compass grass (Figure 2-14). The plant is also known as marram grass. Beach grass can be identified by the upper surface of its leaves, which have ten or twelve parallel veins (raised lines) running from the base to the tip. The plant grows to over two feet in height. Its green and inconspicuous flowers bloom in July

A

B

Figure 2-14. Beach grass (*Ammophila breviligulata*) adds beauty to the beach and stability to the dunes (A). An enlarged seed plume (B).

and August. Fountains of grass are spread by rhizomes, underground runners (stems). When a wind-driven grain of sand strikes a blade of beach grass, it falls down, adding to the dune. The vast network of rhizomes, roots, and erect stems of the plant stabilizes and holds the dune in place, and should never be walked upon. Human feet penetrate the loose sand and damage the rhizomes, causing erosion. Beach grass is so effective at stabilizing dunes that it is often planted in dune-restoration projects (Figure 2-15).

Like beach grass, sea oats (*Uniola paniculata*), also known as southern beach grass, play an important part in anchoring the sand dunes along the Atlantic Seaboard from Virginia to Florida. Wild relatives of the cultivated grain, the stout grass is usually found on the tops of the frontal dunes, growing in dense clumps, with distinctive stalks over five feet in height that are topped with distinctive tawny seed plumes (Figure 2-16). Seeds mature by October and are eaten by birds, deer, rabbits, and rodents.

Poison ivy (*Rhus radicans*) can grow in sand. It is a vine or shrub with three broad (usually smooth-edged) leaflets that turn brilliant red

Figure 2-15. A dune-restoration project using beach grass (*Ammophila breviligulata*) to stabilize the sand.

Figure 2-16. Sea oats (*Uniola paniculata*). The seeds of the distinctive tawny plume are all on one side of the roundish stem (A). Enlarged seeds (B). (Photo courtesy of the National Park Service.)

in the fall (Color Plate 2a). The small, greenish-yellow flowers are in bloom from June through August, with white or grayish-white berries in the fall. "Leaves of 3, let 'em be!" Contact with the leaves, stems and fruit may produce an irritating itchy rash. The plant's tissues contain urishol, an oil similar to carbolic acid, which is responsible for the irritation of human skin. Deer feed on the leaves and vines and birds eat the berries. When recognized by beachcombers, the plant is a great deterrent to walking on the dunes. However, poison ivy is not a deterrent to beach vehicles, which are much more destructive.

Other small dune plants include the seaside goldenrod (*Solidago sempervirens,* Figure 2-17), which can be identified by its golden yellow flowers when in bloom (August to the first frost), and smooth,

Figure 2-17. Seaside goldenrod (*Solidago sempervirens*) with distinctive golden yellow flowers.

broad, fleshy, green leaves. The graceful, plumelike, flowery head is more prominent on plants from Long Island northward, less showy and narrower leaves on plants south of Long Island. Plant size ranges from two to eight feet tall.

The beach pea (*Lathyrus maritimus*), a relative of the sweet pea, has sprawling stems, clusters of purplish flowers, blooms throughout the summer, and produces seed pods (Figure 2-18).

Beach heather (*Hudsonia tomentosa*) is a low shrub that grows to about nine inches in height, and is found from Maine to Virginia. The foliage is grayish-green, with small needlelike leaves, and small, bright yellow flowers that bloom in late spring.

Inland from the beach grass community, shrubs thrive in the sand and add stability and beauty to the dunes. The salt-spray rose (*Rosa rugosa*) is covered with pink or white flowers from late spring to the first frost and produces scarlet rose hips (fruit) (Figure 2-19 and Color Plate 2a). The large hips are made into jelly. The dense shrub grows to over five feet in height. There are five or nine leaflets on a stem. It is

Figure 2-18. Beach Pea (*Lathyrus maritimus*). The specimen is 8 inches high.

Figure 2-19. In late fall, a salt-spray rose (*Rosa rugosa*) is covered with scarlet rose hips. Fountains of beach grass (*Ammophila breviligulata*) surround the small shrub and a seaside goldenrod (*Solidago sempervirens*) is directly behind it. (See also Color Plate 2a.)

uncertain how the plant that was brought to America from Japan in 1872 found its way to the dune community.

Prunus maritima, the beach plum, has dense clusters of small white flowers that may become pink with age and look like wild roses. The flowers bloom from late April through June. The plant sometimes grows sprawling on the ground rather than upright, and is found from Maine to Virginia. When erect, it can grow to five feet in height. The velvety bark is reddish-brown with tan dots. At summer's end, the edible, small, plumlike fruit ripens, but has a sour taste. However, the purplish fruit, like sour cherries, is excellent for jam and jelly.

Another shrub is the bayberry (*Myrica pennsylvanica*). The narrow, aromatic leaves can be used in place of bay leaves, especially in cooking fish dishes. The plant can be identified by crushing a leaf, which gives off a spicy odor. The green flowers bloom in May and June. In the fall, the bush, which grows to more than five feet high, is covered with clusters of tiny, bluish-gray, waxy berries (Figure 2-20). The colonists used the berries to make candles. Bayberry candles and beach plum jam are often found in seaside gift shops.

Seashore plants are often conditioned to endure the harsh desertlike environment. The unshaded sun relentlessly bakes the plants and they must protect against loss of life-sustaining water. The seaside goldenrod has fleshy leaves with a wax coating, which helps retain moisture.

Figure 2-20. In the fall, bayberry shrubs (*Myrica pennsylvanica*) produce clusters of small, light gray berries.

On extremely hot and sunny days the leaves of the beach pea fold up to reduce evaporation. Under the same conditions the long leaves of beach grass curl up like tubes. The hairlike fibers on the leaves of dusty miller not only insulate against the heat, but capture harmful salt before it touches the leaf surface.

Numerous animals are present in the dune environment, including birds feeding on berries. Remember, within the boundaries of a national seashore every living animal is protected, even snakes. Only walk on designated trails, and if you are lucky you might encounter a hognose (*Heterodon platyrhinos*) snake. Unfortunately, most people would react by trying to kill the harmless reptile. The small non-poisonous snake primarily preys on Fowler's toad (*Bufo woodhousei*), another inhabitant of the dunes. It is amusing to watch the hognose go through its routine of defensive posturing when an intruder approaches. The snake's extraordinary behavior in the face of danger often begins with inflating the front part of its body, spreading its neck like a cobra, opening its mouth, and hissing menacingly. This defensive action gives rise to the common name puff adder. If threatening actions are not successful in turning the intruder away, the snake may play dead by rolling over onto its back. Usually, if the hognose attempts this ruse you can pick it up and right it, and it will flop back over. But be sure you are playing with a hognose and not a poisonous snake. The hognose has a prominent upturned snout, a specialization developed for burrowing. It has a short, thick body and the tail is often held in a flat coil.

The zone behind the beach is pleasing to the eye with the undulating sand dunes, the colorful flowers and fruit, and beach grass waving in the breeze. The dunes, however, are very fragile and easily damaged by a person walking on them. Always use designated walkways and boardwalks (when present) to cross the dunes. Conserve our coastal environment.

The construction of homes and beach resorts on the dunes interrupts and alters the natural buildup and stabilization of the sand and leads to rapid erosion of the fragile environment.

COMPOSITION OF THE BEACH

The nature and size of beach material are related to the source of the deposits and the continuing forces that are active on the beach. Sand bottoms are unstable; they constantly shift in response to currents, tides, and waves. Coarseness of the sand and the slope of the beach are both increased by strong wave action. Water movement continuously sorts sand particles, and deposits them by size and density.

The color of a beach reflects its composition. Quartz and feldspar are the ingredients of a white beach. Pink streaks of sand contain garnet, a semi-precious gemstone, and/or rosy quartz. A black beach contains magnetite, a black, iron-based mineral. A small hand magnet swept over it will attract pieces of the mineral to its surface. Thomas Edison mined magnetite on the beach of Fire Island, New York, using an electromagnet, but found it was not economically feasible.

The sand composition of beaches in different regions of the study area covered by this guide will vary because the sand grains have eroded from different kinds of rock.

The beach is a very dynamic environment; changing every day with the movement of sand caused by ebbing and flooding tides. The rather subtle daily changes result in major seasonal changes. Stormy weather, usually during the fall and winter months, produces incoming short-period swells and waves that erode beach sand and carry it offshore. The high-energy waves of winter months tend to remove quartz and feldspar and leave heavier iron-containing minerals. That is why beaches often appear darker in winter than in summer. During the calmer weather of summer, beaches are swept by long-period swells and waves that tend to move the sand back, replenishing the beach. Visit the beach in the early fall and you will see the beach at its maximum buildup. In the spring you will see it eroded away.

Wrack Line

Incoming waves are more forceful than the return flow off the beach. They tear marine life such as seaweeds (algae) from their substrate and carry them high on the beach, where they remain as the water recedes. Seaweeds that are commonly observed may be green, or various shades of brown and red (representative species are listed in other habitats such as rock beaches and bays). All possess the photosynthetic green pigment chlorophyll, but those that are brown or red also have one or more other pigments that provide their characteristic colors. The seaweed and debris at the high tide line is often referred to as wrack. The wrack line may contain straw from salt marsh grass and human flotsam. The masses of dead plant material contain an abundant and varied display of animal life, both marine and terrestrial. Spiders dart around the debris and audacious beach flies buzz over it, then deposit their larvae to develop in the decomposing seaweeds. A disturbed pile of seaweed reveals a host of millipedes, beetles, and amphipods.

Beach hoppers (*Orchestia* species), also called beach fleas, are frequently found under damp seaweed and rocks. The common name reflects the bounding locomotion this amphipod achieves by quickly kicking its three pairs of jumping appendages. Their bodies (¾-inch) are flattened sideways, which makes them easy to distinguish from other crustaceans. They have large, compound eyes and two pairs of antennae; the first pair is very short (Figure 2-21). Beach hoppers burrow an inch or so into the sand during the day, but occasionally come up to follow the receding tide down the beach. They are most active after sunset, but if you move seaweed aside you might see them hop about like the fleas for which they have been named. Beach hoppers feed on the decomposing seaweed and other animals. In turn they are preyed upon by sandpipers, sanderlings, and other shore birds.

It is not uncommon to find clam shells (or the shells of other bivalves) with holes having beveled sides. The holes were produced by carnivorous snails that use their tooth-covered, tongue-like radula to drill through the shells of other mollusks. The radula is then used to tear out small bits of tissue (Figure 2-22).

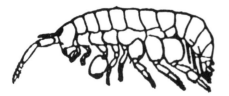

Figure 2-21. The beach hopper (*Orchestia* sp.) is an amphipod crustacean that uses three pairs of jumping appendages to leap into the air.

Figure 2-22. The hole in this bivalve mollusk shell was made by the radula of a gastropod mollusk.

A gastropod shell that occasionally washes ashore from Cape Hatteras southward, but usually not north of the cape, is the Scotch bonnet (*Phalium granulatum*). The egg-shaped shell has deeply incised transverse lines, is white in color with brown rectangular spots, and grows to over three inches. The brown spots may be faded in specimens that have been washed ashore. In 1965, the North Carolina Assembly designated the attractive shell as the state's seashell (Figure 2-23).

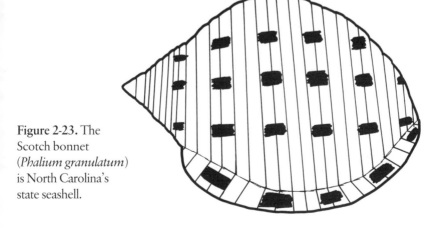

Figure 2-23. The Scotch bonnet (*Phalium granulatum*) is North Carolina's state seashell.

Vermicularia spirata produces an unusual looking snail (gastropod) shell. As the animal matures, the shell it continues to build develops separated whorls that give it a corkscrew appearance; it can be five inches long. The shell looks more like that of an annelid tube worm than that of a gastropod. The common name worm shell denotes this. However, the shell's closed end still shows the spiral gastropod character. The shells are usually found attached to rocks and other shells, but they occasionally wash ashore (Figure 2-24).

An unusual looking shell occasionally found from Cape Cod southward, is the ram's horn (*Spirula spirula*). It has the appearance of a coiled tube worm (annelid), but actually is the shell of a squidlike cephalopod mollusk (Figure 2-25). The thin, snow white, tube-like shell, which grows to one inch in diameter, is composed of chambers of increasing size. The septum (partition) nearest the opening has a small hole in the center (the siphuncle, a tube within the larger tube, connects the chambers). The coils of the fragile shell are separated. The internal shell chambers may serve as a buoyancy device. The mollusk has eight short arms and two long tentacles, like a squid, and lives in deep water. After the animal dies, its small shell may float ashore.

Figure 2-24. The shell of the gastropod *Vermicularia spirata* has the appearance of an annelid tube worm more than that of a gastropod mollusk. The common name worm shell denotes the resemblance.

Figure 2-25. The shell of the ram's horn (*Spirula spirula*) resembles that of an annelid tube worm, but actually is the shell of a cephalopod mollusk.

Cliona species of sponges are called boring sponges because they bore extensive thin, branching tunnels into empty mollusk shells. The shell protects them from predators. Specialized cells secrete an acid that dissolves the shell. They are not parasitic, but the tunnels weaken the shell. The sponge is usually sulfur-yellow in color, and is visible where the bored channels reach the surface. Different species can be identified by the size of their bored holes. In one species, the sponge eventually grows over the host shell, dissolving it, and becomes a massive "free-living" form (Figure 2-26).

Figure 2-26. Boring sponges (*Cliona* sp.) penetrate clam and oyster shells such as the 4.25-inch oyster shell shown here, riddling them with tunnels and openings. The disintegrating shells contribute to the composition of the sand.

Suberites ficus, a sponge commonly referred to as fig sponge, is usually found attached to bivalve shells at depths of 50 feet or more. Occasionally, one will break off and be washed ashore. The fig sponge is usually oval shaped, several inches long, and about ¼ inch thick. It has a shiny appearance and its texture is similar to smooth cardboard, with few and inconspicuous openings. The fig sponge is found from the Arctic to Delaware. In Block Island Sound, the sponge creates a nuisance for commercial fishermen by clogging their nets. Due to its shape and yellow color, the fishermen contemptuously refer to it as elephant dung.

Haliclona oculata is another sponge that is often washed ashore and found in the wrack. It is commonly known as dead man's fingers, because of its fingerlike lobes reaching up from the bottom. The lobes are supported by a short, slender stalk. It may be highly branched and more than 12 inches in length. Oscula, excurrent openings that allow water to pass out of the cavity within the sponge's body, are conspicuous and scattered over the surface (Color Plate 3b). Color ranges from yellow to light brown, occasionally purple or orange tinted. Specimens washed up on the beach and dried may be light gray in color. The sponge is found more frequently in the northern part of the study area.

If large pieces of sandstone, peat, or clay are seen on the beach, boring clams, the angel wing (*Cyrtopleura costata*) and false angel wing (*Petricola pholadiformis*) may be found inside by carefully breaking the material apart. The beautiful, fragile, chalky-white shells are found throughout the study area, especially in marshes. However, the angel wing is more abundant in the northern region, the false angel wing from Cape Cod south but most prevalent below Long Island (Figure 2-27). The elongate oval shell has filelike rasping areas on the anterior part that are used to burrow into the substrate as the animal hangs on with its broad, muscular foot. The thin shell of the angel wing grows to more

Figure 2-27. The valves of a 1.75-inch-long false angel wing (*Petricola pholadiformis*).

than eight inches long, but only grows to about two inches for the false angel wing. Boring clams can be found in the sandy mud, clay, or peat around the low-tide mark in other habitats, such as bays and estuaries.

The hydrozoan *Physalia physalia* is commonly known as Portuguese man-of-war. It is easily recognized by the large rainbow-hued, bladderlike, gas-filled float (Figure 2-28). The animal cannot swim and drifts along at the mercy of currents and wind. The float can be twelve inches in length and the extended tentacles can trail out 75 feet or more. The tentacles, like those of jellyfish, are translucent and hard to see by bathers. The neurotoxin secreted by the nematocyst (stinging structure) can cause severe pain and shock, and even respiratory paralysis. A woman stung by a Portuguese man-of-war in the waters off Hyannis, Massachusetts, was driven from a mild to a severe reaction with shock by the casual application of cold, wet (freshwater) towels. The pain was relieved instantaneously with isopropyl (rubbing) alcohol sponging. However, some Australian researchers have watched cnidoblasts under the microscope fire off nematocysts during application of alcohol, while they are paralyzed by acetic acid (vinegar). Also, meat tenderizer applied to the skin, reportedly will break down the toxin's protein. To add to the confusion on how to treat a sting, "lifeguards in Florida report that a brisk freshwater shower controls the sting of certain jellyfish, while it worsens the sting of others." If the symptoms are severe, a physician should attend the victim. Fortunately, *Physalia* is not frequently found north of Cape Hatteras. Occasionally, one is washed up on the beach, especially after storms have blown them in from the Gulf Stream; care should be used if an attempt is made to inspect one. The float can be handled, but the tentacles made up of many polyps can still discharge nematocysts. If stung, one should carefully remove the tentacles and shave the affected area.

Physalia is a large colony of polyp types that integrate to function as one individual. There may be up to 1,000 individuals with intercommunicating gastrovascular cavities in a single colony. Some polyps are specialized for capture of prey, usually fish, while others are adapted for digestion or reproduction.

A fertilized egg develops into a single larval polyp, which matures into a float. A colony is produced when the float produces zones of budding, which develop the other members of the colony. One species of

Figure 2-28. Portuguese man-of-war (*Physalia physalia*).

fish (*Nomeus gronovii*) is not affected by *Physalia*'s tentacles and swims freely under the colony. This is an example of symbiosis, where the small fish acts as a lure, drawing a larger fish within range of the tentacles. The small fish benefits, in turn, by the protection offered by the tentacles and the fragments of food regurgitated by *Physalia*.

Surprisingly, considering its venomous stings, some animals prey upon *Physalia,* including sea turtles, at least one species of fish, and a small snail, *Janthina janthina.* The snail produces a raft of gelatinous bubbles and drifts upside down with the Gulf Stream. When the snail comes into contact with a Portuguese man-of-war, it abandons the bubble raft and feeds on the hydroid, then secretes another raft and drifts away. The delicate, almost translucent, violet-colored shells of *Janthina*, about an inch long, occasionally wash ashore. The shell color gives rise to the common name violet snail.

Clear gelatinous blobs lying on the sand may be the remains of jellyfish.

The tests (skeletons) of dead sand dollars are occasionally found on the beach. There are two species of this cookie-shaped echinoderm common to the study area. The keyhole sand dollar (*Mellita quinquies-*

perforata) is more frequently encountered south of Long Island, while the Atlantic sand dollar (*Echinarachnius parma*) is found in greater numbers north of Long Island. Both species possess the distinctive star pattern of small openings that allow the tube feet to protrude. The keyhole sand dollar can be easily distinguished by the five slitlike holes that give rise to the common name (Figure 2-29). Both species grow to about three inches in diameter. While alive, sand dollars are covered with small brown spines, that gives them a velvety appearance (Color Plate 13c). Tube feet (part of the water vascular system described in Chapter 1) gather detritus for food as the animal moves over the bottom. Detritus is decomposing plant and animal matter. Detritus is a French term meaning, "disintegrated matter." The food is channeled along "valley-like grooves" on the undersurface (oral surface) toward the mouth in the center. Five sets of teeth and sieve plates inside the mouth are jointly called Aristotle's lantern (Figure 3-6). Aristotle is so honored because his *Historia Animalium*, the first great zoology textbook, describes the structure. Sand dollars are preyed upon by sea stars and fish.

Like sea urchins, the dead sand dollars usually lose their spines as they wash ashore. The madreporite, which allows water to enter the water vascular system, is in the center of the star pattern on the top (aboral surface). Five small openings around the madreporite allow

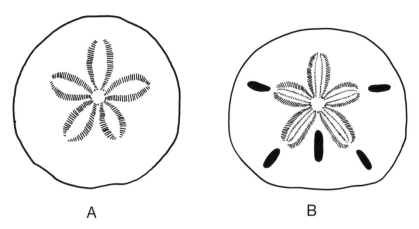

A   B

Figure 2-29. The Atlantic sand dollar (*Echinarachnius parma*) (A). The keyhole sand dollar (*Mellita quinquiesperforata*) (B).

sperm and eggs to pass out into the surrounding water where fertilization occurs. If a dead sand dollar is gently shaken, the separated teeth of Aristotle's lantern may produce a rattling sound. Dead sand dollars can be bleached white by soaking overnight in a solution of one part clorox to three parts water. Rinse off and place them outside in the sun to dry and finish bleaching. Collect only dead sand dollars and leave some for other beachcombers.

The flat, leatherlike, black egg cases of skates (*Raja* species, a cartilaginous fish) frequently wash ashore. The egg case looks like a small rectangular coin purse with long, curly tendrils at the four corners; it is commonly known as the mermaid's purse, or sailor's purse (Figure 2-30). Skates lay their egg cases in the quiet water of bays and the four curly tendrils help to entrap them in eel grass to prevent them from washing ashore. The hatchling looks like a miniature adult, and if the egg case is broken open, the young skate may have escaped. If the case is unopened, the embryo, or its remains, is probably still inside.

The egg cases of the dogfish shark (*Squalus acanthias*) occasionally wash ashore. They are similar in appearance to skate egg cases, but are creamy in color and the tendrils are highly curled.

Figure 2-30. The egg cases of skates (*Raja* spp.) frequently wash ashore. The specimen is 6 inches from tendril tip to tendril tip.

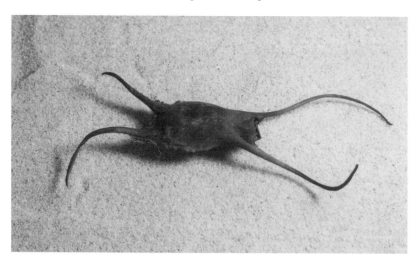

Plate 1a. Westhampton barrier island, Long Island, New York. The Atlantic Ocean on the left, Moriches Bay on the right. Moriches Inlet and Fire Island can be seen in the distance.

Plate 1b. A salt marsh at Fire Island encroaches into the bay. Note the mosquito control ditches.

Plate 1c. An extensive mudflat in Shinnecock Bay, Long Island, is exposed at low tide.

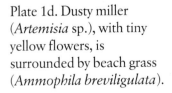

Plate 1d. Dusty miller (*Artemisia* sp.), with tiny yellow flowers, is surrounded by beach grass (*Ammophila breviligulata*).

Plate 2a. Pink flowers of the salt spray rose (*Rosa rugosa*). Note the intermingled poison ivy (*Rhus radicans*). The old adage "leaflets three, let it be," will spare you much pain and grief.

Plate 2b. *Laminaria* sp. with the white hydroid *Obelia* growing on one plant. Note the sea star (*Asterias forbesi*) that has lost two arms. (Photo by Norman Despres.)

Plate 2c. An alga is surrounded by a sponge, note the volcano-like oscula, and almost completely covered by a bryozoan (*Membranipora* sp.). (Photo by Jim Matulis.)

Plate 2d. The encrusting calcareous red alga *Phymatolithon laevigatum* shares a rock with the hydroid *Tubularia* and green sea urchins (*Strongylocentrotus droebachiensis*) that feed on the alga. (Photo by Norman Despres.)

Plate 3a. The green alga sea hair (*Enteromorpha* sp.) and young salt marsh grass (*Spartina alterniflora*).

Plate 3b. Dead man's fingers sponge (*Haliclona oculata*). Note the oscula. (Photo by Dave and Sue Millhouser.)

Plate 3c. The colonial hydroid *Tubularia crocea*. (Photo by Dave and Sue Millhouser.)

Plate 3d. The pink-hearted hydroid (*Tubularia*) has two widely separated rings of tentacles. Clusters of grape-like medusae can be seen between the tentacles. They remain attached to the parent polyp. (Photo by Dave and Sue Millhouser.)

Plate 3

Plate 4a. Lion's mane jellyfish (*Cyabea capillata*). (Photo by Norman Despres.)

Plate 4b. Frilled sea anemone (*Metridium senile*). (Photo by Dave and Sue Millhouser.)

Plate 4c. Expanded and contracted sea anemones (*M. senile*). With tentacles and column contracted, they look like small volcanoes. (Photo by Dave and Sue Millhouser.)

Plate 4d. Tentacles surround the orange-ringed mouth of a frilled sea anemone. (Photo by Norman Despres.)

Plate 4

Plate 5a. Sea gooseberry comb jelly (*Pleurobranchia pileus*). (Photo by Dave and Sue Millhouser.)

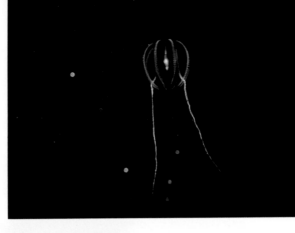

Plate 5b. Plumed sea slug (*Aeolidia papillosa*). (Photo by Norman Despres.)

Plate 5c. Red-fingered sea slug (*Coryphella rufibranchialis*). (Photo by Norman Despres.)

Plate 6a. The tortoise shell limpet (*Acmea testudinalis*) blends well with the background. (Photo by Norman Despres.)

Plate 6b. The moon shell (*Lunatia heros*) glides over the bottom on its huge foot. Note the sand dollar (*Echinarachnius parma*) and the long-clawed hermit crab (*Pagurus longicarpus*) above the snail. (Photo by Norman Despres.)

Plate 6c. The common periwinkle (*Littorina littorea*) on the stipe of a kelp. Note the red alga (*Phymatolithon laevigatum*) on the shell's spire. (Photo by Norman Despres.)

Plate 7a. A waved whelk (*Buccinum undatum*) has difficulty in finding bare rock among green sea urchins (*Strongylocentrotus droebachiensis*). (Photo by Norman Despres.)

Plate 7b. Waved whelk eggs. (Photo by Dave and Sue Millhouser.)

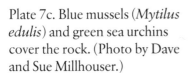

Plate 7c. Blue mussels (*Mytilus edulis*) and green sea urchins cover the rock. (Photo by Dave and Sue Millhouser.)

Plate 7d. Feeding blue mussels with open valves. (Photo by Jim Matulis.)

Plate 7

Plate 8a. Ribbed mussels (Modioleus demissus), two-thirds buried in the sediment, have prominent ribs on the valves. Note the acorn barnacles (*Balanus* sp.) and the blue mussel valve on the left (no ribs).

Plate 8b. The siphons of an almost completely buried clam take in water for feeding and respiration. (Photo by Jim Matulis.)

Plate 8c. A squid (*Loligo brevis*) blends well with the sand bottom. (Photo by Dave and Sue Millhouser.)

Plate 8d. A green sea urchin (*Strongylocentrotus droebachiensis*) and a mop of squid eggs. (Photo by Dave and Sue Millhouser.)

The significance of one type of debris occasionally encountered by a beachcomber is often overlooked. Pieces of coal on the beach often indicate an offshore shipwreck. The remains of wrecked ships abound along America's Eastern Seaboard. Thousands of vessels have sunk along that shore since colonial times—victims of winter storms, dense fog, military conflict, human error, inhospitable shores. The remains of some shipwrecks buried in the sand appear and disappear with the wind and tides as they are alternately exposed and buried by the ever-shifting sand. Some offshore wrecks may extend above the water (at least at low-tide). However, most are completely submerged and only an occasional reminder of their presence washes ashore in the form of pieces of coal or timbers with bronze fastenings protruding (Figures 2-31 and 2-32).

Cape Hatteras is known as the "Graveyard of the Atlantic," a name used by Alexander Hamilton, who as a young man sailed past the area. Later, as secretary of the treasury, he used his influence to have a light-house built there. Cape Hatteras earned its reputation as a dangerous area for shipping because of shifting sand bars, strong currents, and sudden storms (Figure 2-33).

Figure 2-31. Pieces of coal on the beach at East Hampton, Long Island probably indicate an offshore shipwreck.

Figure 2-32. A large piece of wreckage washed ashore at Westhampton Beach, Long Island, after a severe winter storm.

Figure 2-33. The wooden-hulled side-wheel steamer *Pocahontas* foundered off Cape Hatteras in 1862. The National Park Service believes the upper part of a steam engine that is exposed off Sand Street, Salvo, N.C., although locally said to be from the steamer *Richmond*, is the remains of the *Pocahontas*.

A Boston newspaper of the 1800s called Cape Cod "our sailors' graveyard." Jutting out into the major shipping lane between New York and Boston, the arm of sand and gravel has been responsible for the loss of an estimated 3,000 ships. Again, dangerous shoals and unpredictable storms claimed many ships, but dense fog produced by the meeting of the cold Labrador Current and warm Gulf Stream was responsible for most of the strandings and collisions.

Pieces of driftwood often assume intriguing shapes after they have been scoured by sea and sand. The wood is often carrying passengers such as seaweed and acorn barnacles and goose neck barnacles. There is a fish commonly found sheltering underneath drifting timber. It is appropriately named the wreck fish. Driftwood may be riddled with tunnels by marine borers such as the shipworm and gribble.

The most productive time to explore the wrack line is after a storm, the more severe the better. The waves uproot plants and animals and wash them up on the shore to die, to the delight of the ghost crab, beach hoppers, and the shore birds, not to mention the beachcomber. Also, the fresh organisms will be in a better state of preservation. Subtidal seaweeds such as *Laminaria* may have been torn from their substrate and washed ashore. Check the plants for epiphytic plants, and animals such as hydroids and bryozoans. Animals like sponges and sea stars might not have been able to withstand the awesome force of the waves and be high and dry.

Shore birds, sand fleas, beetles, and other animals clean the beach by eating the plants and animals of the wrack line. They play an important role in the removal of the decomposing material.

Unfortunately, humans have added trash to the high tide level of the beach. Discarded items, such as styrofoam cups, bottles, and various plastic items tossed into the water from boats and people along the the shore drift back onto the beach. It is not only unsightly, but hazardous to animals that eat it or become entangled in it. Most of us have seen an animal or the photograph of an animal with a plastic 6-pack carrier draped around its neck.

SANDY HABITATS

Attached seaweeds are seldom found in the intertidal zone of an ocean sand beach because wave action keeps the surface layers of sand

in constant motion. Stand still in the intertidal zone and feel the pleasant sensation of sand washing from under your feet. As you walk away,
notice how quickly the footprints disappear. The rapidly moving sand
eliminates any stable substrate, other than an occasional rock for attachments. The beach appears lifeless, but there are beach animals; they are
almost exclusively burrowers, and remain permanently or temporarily
below the surface, to avoid drying out when the tide recedes. Any beachcomber who has built a sandcastle knows that the sand, even in the higher intertidal zone, is moist only a few inches below the surface.

The ghost crab (*Ocypode quadrata*), also called racing crab or sand
crab, lives in a burrow well above the high-water mark. Its tunnels can
extend down four feet; the older crabs have burrows in the dunes, the
burrows of young crabs are closer to the water. Burrow openings are
approximately two inches in diameter and about three or four feet apart.
A ghost crab is dormant in its burrow during the winter. It is well adapted to such a dry area, and only occasionally enters the water to damp-
en its gills and partially fill a small chamber that is part of the respiratory system. The female crab deposits eggs in the ocean, and after
hatching, larvae live there until instinct urges them to move onto shore.
The ghost crab has a square carapace and long eye-stalks (Figure 2-34).
The large eye at the end of an eye-stalk gives the appearance of a
periscope (Color Plate 11b). The rotation of the eye-stalks gives the

Figure 2-34. The swift-footed ghost crab (*Ocypode quadrata*) occasionally is
found as far north as Rhode Island.

animal 360° vision. When the crab enters the burrow, the long eye-stalks recess into grooves on its shell (Color Plate 11c). It runs side-ways on the tips of its toes and is very fast, thus the common name rac-ing crab. The genus name *Ocypoda* means swift-footed. At an estimated speed of 10 miles per hour, they are difficult for the human eye to follow. Groups often run with military precision. They travel faster than a human, and can burrow into sand in less than one second. The best way to observe them is through binoculars. But their sand-like coloration makes them almost invisible when they are stationary; thus another common name, ghost crab.

These crabs are formidable predators in sand dunes and the intertidal zone, usually feeding at night on beach fleas, mole crabs, and coquina clams. Herring gulls prey upon ghost crabs. The crabs usually venture out after sundown to forage, after the gulls have left the beach. At night, in the southern region of the study area, the beach is alive with ghost crabs. Walk the beach with a flashlight and you will see all sizes. A ghost crab will remain stationary in a beam of light for a few moments, then just vanish. Although the crab is more active at night, it is also seen in the day, especially on dark, overcast days. It is found on less populated, undisturbed beaches from Virginia southward, but occa-sionally as far north as Rhode Island.

Just above the high tide mark look for pencil-size, oval holes. Sev-eral inches below the surface shrimplike sand fleas (*Talorchestia* species) may be found. They burrow into the sand as the tide advances and come up to forage as the water recedes. These sand-colored amphi-pod crustaceans are slightly larger (to one inch) and lighter in color than beach fleas (*Orchestia* species) which are found under damp sea-weed and other debris (Figure 2-35).

Just below the high tide mark, you may see small holes in the sand. They appear to have been produced by burrowing animals. However, most of the openings are "percolation holes," caused when a wave washes over the sand, and air trapped in the sand escapes upward.

Few animals and fewer plants can survive in the intertidal zone where the sand is unstable and abrasive. At high tide the sand is con-stantly shifting, while at low tide it is hard-packed. There may be only a few species in this harsh environment, sometimes as few as six in some areas along the south shore of Long Island, but those that have

Figure 2-35. The sand flea (*Talorchestia* sp.), like the beach flea, uses large posterior appendages to hop along the beach.

adapted to its stresses may reach enormous numbers. Beachcombing on the sand beach and mud flats of bays and estuaries will produce a greater variety of organisms.

A few species of crustaceans are found in the intertidal zone of an ocean beach. Small isopods (*Eurydice* species), less than one-half-inch long, burrow into the sand and feed on microscopic animals. Their bodies are dorsoventrally flattened and the thorax has seven segments, the abdomen six. The mole crab (*Emerita talpoida*), also known as the sand mole, is a crustacean but not a true crab. It burrows in the sand tail first with only its eyes and large feathery antennae protruding. The small body is streamlined and some of the appendages are modified to form digging organs for burrowing in the sand (Figure 2-36). Another common name is the sand bug. The body is egg-shaped, with a gray carapace. The mole crab faces incoming waves, and when a wave breaks over the animal and begins to recede, the feathery antennae waving above the sand filter plankton out of the water. Other feeding appendages move the food to its mouth.

When mole crabs are present, they are found in large numbers. They move offshore during the winter. At low tide stand in the water and observe the sand as the wave recedes. You may see a mole crab's feathery antennae extending above the sand. The antennae will disappear with the incoming wave. The animal does not have pincer claws and can be handled safely. Catch one and while gently holding it in one hand note that the posterior legs and telson are modified for digging. Scoop up a handful of wet sand with the free hand and place the mole crab on top of the sand, then watch it dig downward toward your palm.

Figure 2-36. The mole crab (*Emerita talpoida*) has a small body (1½ inches in this specimen) that is streamlined for burrowing in the sand.

The female mole crab can be more than 1½ inches long, but the male is much smaller. Males are eaten by sanderlings, but the females are too large. The blue crab uses a unique method to catch a mole crab. It grabs the much smaller animal with one pincer claw and runs in a circle until the mole crab pops out of the tightly packed sand. Mole crabs are used as bait by surf fishermen, especially if the female is carrying a large mass of orange eggs.

The calico crab (*Ovalipes ocellatus*) is a swimming crab, but often, like the blue crab, it will bury itself in the sand with only the eyestalks exposed. The shell shape is circular and is light bluish-gray in color with purple spots, resembling a calico pattern. Most of the crab shells washed ashore are casts from molts, are very light in weight, and have no offensive odor of decomposing flesh.

Oxyurostylis smithi, another crustacean, is commonly known as the sharp-tailed cumacean because its telson is tapered to a sharp point (Figure 2-37). The small animal grows to about one-quarter-inch in length and is usually found in the sand.

Figure 2-37. The sharp-tailed cumacean (*Oxyurostylis smithi*) has a telson that is tapered to a sharp point.

Some mollusks, including the bivalve coquina (*Donax* species), also called digger wedge shell, are found in the lower region of the intertidal zone. Coquina grows to about three-quarters of an inch; its shell color may be white, yellow, red, or olive, often with colored rays that provide a variable rainbow spectrum of colors (Figure 2-38). Internal shell coloration may not be the same as external colors. Although such a substrate would seem to offer an extremely difficult habitat for a bivalve, it is usually found in the hard sand beaches that receive the full impact of the waves from the open ocean. The animal is more numerous south of Cape Hatteras, but is encountered as far north as Long Island.

Coquina allow wave energy to carry them up and down the beach face. When the tide begins to flood, agitation by the incoming waves stimulates them to literally jump out of the sand, and be carried along with the water. They trail a muscular foot like an anchor until the wave begins to recede. Then, the foot quickly burrows into the sand until another wave provides another free ride. The process is reversed during an ebb tide; they emerge during the back wash to be carried down the beach face, again expending little energy. Because they respond to the weight of waves, stamp heavily on the sand and you may trick one into popping up prematurely.

Catch a coquina and scoop up a handful of wet sand with the free hand. Place the bivalve on top of the sand and watch it extend its mus-

Figure 2-38. The coquina clam (*Donax* sp.) is found in the harsh environment of the intertidal zone.

cular foot and pull itself down into the sand with spasmodic jerks. The small bivalve is often found in enormous numbers and, like the mole crab, it is boiled (in large numbers) to make a broth for a chowder.

Severe storms can dramatically deplete a beach, not only of sand, but also an entire population of coquina and other animals such as the mole crab. It may take years for the animals to reach pre-storm numbers.

The largest of bivalves found in the study area is *Spisula solidissima*, up to seven inches in length, commonly known as surf clam, skimmer, hen clam, or sea clam. As one of the common names implies, the bivalve is found in sand in the surf zone, in deeper waters, and is the most common clam shell found on ocean beaches south of Cape Cod. The valves are thick, but thinner than the quahog, somewhat triangular-shaped, and have a large centrally located umbo, while that of the quahog is off to one side. There is a large spoon-shaped cavity for the ligament in the hinge area (Figure 2-39). The valves are smooth, with fine growth rings and thin, brown periostracum. Color is yellowish white.

Figure 2-39. The surf clam (*Spisula solidissima*) is the largest bivalve found in the study area. The specimen is 5.25 inches long. The thick valves are somewhat triangular-shaped and have a centrally-located umbo. Note in the hinge area, the large spoon-shaped cavity for the ligament.

The commercially important clam is very popular for clambakes, but only the adductor muscles are eaten.

TURTLES

Several species of marine turtles are found not only in the ocean but in the bays and sounds during the warm months, although few of us ever see them. Most sightings occur when turtles wash ashore entangled in fishermen's nets, or in the southern region of the study area, when laying eggs in the sand. If you happen upon a sea turtle nesting on the beach at night, do not shine a flashlight on the animal, they are easily disturbed.

The shells of sea turtles are usually heart-shaped and their legs are flippers. Examples of sea turtles found in the study area are the hawksbill (*Eretmochelys imbricata*) and the leatherback (*Dermochelys coriacea*). The hawksbill prefers warmer waters and is more frequently seen in the Caribbean, but is found as far north as Cape Cod. The common name is derived from the beak-like upper jaw that overhangs the lower jaw (Figure 2-40). This endangered species was once hunted for its beautiful shell that was used to produce "tortoise shell" jewelry. The leatherback is easily distinguished from the hawksbill and other sea turtles by its ridged leathery back (Figure 2-41). Instead of horny plates, its outside covering is a leathery skin. The horny shell consists of platelets of bone embedded in the skin. Unlike all other turtles, it cannot retract its head and limbs. The flippers do not have claws. It is the largest tur-

Figure 2-40. The hawksbill sea turtle (*Eretmochelys imbricata*) has a beak-like upper jaw that overhangs the lower jaw.

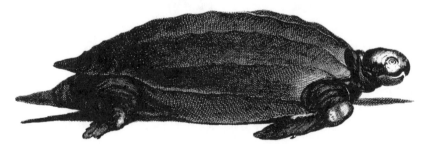

Figure 2-41. The leatherback (*Dermochelys coriacea*) is the largest sea turtle and the only one without a true shell. Five to seven ridges are on the carapace.

tle found in the study area—growing to more than ten feet and weighing up to a ton—and feeds primarily on jellyfish. The leatherback turtle has been found as far north as Nova Scotia.

Organizations like the Okeanos Ocean Research Foundation on Long Island record the sightings of sea turtles and mammals (e.g., whales and seals) and rescue and rehabilitate injured or sick animals through their stranding programs. Sightings and strandings should be reported to:

Delaware—Division of Fish & Wildlife (302) 739-4782

Connecticut—Mystic Aquarium (203) 536-9631

Maryland—National Aquarium (410) 576-3853

Massachusetts—New England Aquarium (617) 973-5273

New Jersey—Marine Mammal Stranding Center (609) 266-0538

New York—Okeanos (516) 728-4522

North Carolina—NMFS Beaufort Lab. (919) 728-3595

Virginia—Marine Science Museum (804) 437-4976

VISIT AN OCEAN BEACH

Unless you are as agile as a sanderling, plan on getting your feet wet while searching for burrowing organisms in the intertidal zone of an ocean beach.

Beach walking and wading is not without its hazards. There is always the chance of stepping on glass or a nail projecting from a plank of wood. Wearing shoes is strongly recommended.

The austere environment of an ocean sand beach does not offer the variety of organisms found on rocky shores, but for the determined, patient beachcomber there is much to see.

3

Rock Beaches
and Jetties

I wiped away the weeds and foam,
I fetched my sea-born treasures home;
But the poor, unsightly, noisome things
Had left their beauty on the shore,
With the sun and the sand and the wild uproar.

Ralph Waldo Emerson

The predominately sand beaches that stretch from Cape Cod to the north shore of Long Island are occasionally interrupted by areas of rock. And there are many manmade rock jetties and breakwaters from Cape Cod to Cape Hatteras (Figure 3-1). The period of exposure to view the rock surfaces ranges from seldom to most of the time.

Tidal fluctuations expose intertidal plants and animals that are attached to the hard, fixed, rocky substrates. Only the hardiest organisms can survive in this harsh environment, where they must deal with extreme temperature fluctuations and exposure to air and potential desiccation. When the tide ebbs, water is often trapped in depressions and cavities in the rocks, forming tide pools; they are ideal for beachcomber

Figure 3-1. A manmade rock jetty protects Moriches Inlet on Long Island and provides a hard substrate for marine organisms to attach to.

exploration. Crustaceans, such as hermit crabs, and echinoderms, such as sea urchins and sea stars (starfish), frequent such areas. Even small fish may be trapped in their shallows as the water recedes. Sit quietly beside a tide pool; do not move and shortly organisms will begin to move about. Turn over rocks lying in shallow water to look for animals attached to the undersurface. Carefully replace the rocks in the same position when finished.

The distribution of the plants and animals in the intertidal area depends on their ability to cope with the environmental extremes of exposure, turbulence, and prolonged periods out of the water as the tide recedes. It is often easy to discern specific bands of marine organisms at low tide. They are called zones, and generally run parallel to the area's coastline. Such zonation is easily recognized on rocky shores because few animals or plants can penetrate their hard surface. The marine life of sand or mud beaches often find a haven underground.

Seaweeds (algae) usually dominate the zonation of an area, with different species growing at distinct levels on the rocks. They have strong holdfasts and flexible, stemlike stipes. Some have bladderlike, gas-filled swellings that allow them to float with the tide, and receive as much light as possible. Some that can withstand considerable battering by the sea can be found on rocks that are exposed to the full force of the waves.

ANIMALS OF ROCKY REALMS

Some of the many animals that feed on seaweed eat fully grown plants; others graze on microscopic sporelings. The chief grazer is the Atlantic plate limpet (*Acmaea testudinalis*), also called the tortoise shell limpet, a single-shelled mollusk. The shell is conical and oval-shaped, with no opening at the top (Figure 3-2, and Color Plates 6a and 12c). It is not coiled like a snail shell and grows to about an inch in length. Shell color varies, but is often a dirty white, bluish white, or yellow, with brown streaks radiating from the apex. The interior surface is bluish with a dark brown center and a brown and white checkered border.

Limpets are found on rocks in the intertidal zone, in tide pools and sublittoral; they possess a homing instinct and return to the same spot on a rock after each feeding sortie, usually no more than three feet. They feed at night, usually at high water, using their muscular "foot" to glide over the rock and scrape algae off with their filelike tongue (radula). Limpets may move about on cloudy days if the rock is wet. They

Figure 3-2. The Atlantic plate limpet (*Acmaea testudinalis*) does not have a hole at the top of the shell, as does the keyhole limpet.

occasionally eat through the base of larger plants, dislodging them from their rock attachment. Grazing on seaweed, they do not compete for food with filter feeders like barnacles. The removal of algae from hard surfaces increases the surface area open to settlement by the barnacles. Limpets use their shells to grind down the surface of a soft rock, creating a depression into which they fit perfectly. On hard rocks the shell's margin is worn away, also resulting in a close association between shell edge and rock surface, but leaving no depression. Their strong, muscular foot makes removal of limpets difficult without damaging the shell. It requires about a 70-pound pull to remove a limpet having a basal area of one square inch. However, if you approach cautiously and apply sudden pressure with your thumb, you may dislodge one. If the limpet had been attached to a soft rock you will probably notice a ring-shaped depression. Put the specimen back in the same spot when finished. The shell's shape and the foot's powerful adhesion allow limpets to withstand heavy seas without dislodging. When a wave strikes the conical shaped shell, it pushes the limpet against the rock rather than washing it off. When the tide is out the margins of shell are clamped down against the rock so tightly that water is retained in the narrow groove around the foot, bathing the gills until the tide returns. *Acmaea* are frequently found from Long Island Sound northward.

Diodora (*Fissurella*) *cayenensis*, found from New Jersey southward, is commonly called the keyhole limpet because of the hole in the top of its shell (Figure 3-3). The conical shell shape with the opening at the top gives the appearance of a miniature volcano. For respiration, water is drawn under the shell near the head and passes out through the opening above. Shell color is dirty white. The larval stage of a limpet settles on a rock after a short free-swimming period.

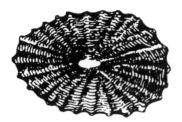

Figure 3-3. The keyhole limpet (*Diodora cayenensis*) is easily distinguished by the hole in the top of its shell.

Chitons, which also graze on seaweeds, are often mistaken for limpets, a gastropod. However, chitons belong to a different class of mollusks, amphineurans. *Chaetopleura apiculata* is an example of a chiton found in the study area covered by this guide. Close inspection reveals that the shell of a chiton is composed of eight separate plates (Figure 3-4). In Britain, they are called coat-of-mail shells. The articulated plates, which are gray, yellowish, or reddish in color, allow the animal's body to conform to the irregularities of the rock surface. The girdle (mantle) is sparsely covered with short hairlike spines. These small chitons, to about one inch in length, are found on rocks low in the intertidal zone or below. Much larger species, three to eight inches long, are found in the tropics. Like limpets, they feed at night and will return to the same spot on the rock. Also, like limpets, chitons are difficult to remove from a rock without harming the animal. However, should you remove a chiton from a rock it will roll up, much like a pillbug. Straighten it out, then turn it over and the foot and head can be seen on the undersurface. The head does not have eyes or tentacles, but is identified as a rounded area, with an opening, the mouth, in the center. Gills lie in the narrow groove that encircles the foot.

Another seaweed feeder is the spiny-skinned sea urchin. The purple sea urchin (*Arbacia punctulata*) is found in the lower intertidal zone throughout the study area. It is usually purple in color as the common

Figure 3-4.
Chaetopleura apiculata, like all chitons, has a shell composed of eight articulated, but separate, plates.

name indicates, but occasionally may be reddish-brown (Color Plate 14c). The test can be about two inches in diameter and the spines about one inch long (Figure 3-5).

The green sea urchin (*Strongylocentrotus droebachiensis*, the longest scientific name of any animal; thankfully there is a common name) is found on the bayside of Cape Cod and northward, seldom on the south shore of Cape Cod. It can easily be distinguished from the purple sea urchin by its color (Color Plates 2d, 7a, 7c, 8d, 14a, and 14b). Also, the test is larger, usually about three inches, and the spines are shorter than the purple sea urchin's. The roe (eggs) of some sea urchins are eaten in Europe and the Orient, and are offered in sushi bars as "uni." Green sea urchins produced a $15 million harvest in the state of Maine in 1992, which has surged since. Once prolific along Maine's coast, green sea urchins are becoming far less common. The roe from harvested sea urchins are sold to Japan.

Sea urchins possess a set of five sharp teeth, called Aristotle's lantern (similar to sand dollars, see Chapter 2), that are used to tear algae into tiny bits for consumption (Figure 3-6).

Acorn barnacles (*Balanus* species) are among the most conspicuous animals on the rocks. They cement themselves firmly in place, high in the intertidal zone, often collecting in such numbers that they form a white band that marks the uppermost tidal level. The external appear-

Figure 3-5. The spiny-skinned purple sea urchin (*Arbacia punctulata*).

Figure 3-6. Aristotle's lantern, five teeth and associated structures, from a sea urchin.

ance of their volcano-like shell makes them look like mollusks, which is how they were classified until 1830. The shell is composed of fused calcareous plates and the opening (orifice) is closed by hinged valves (Color Plates 8a and 9c). Different species are distinguished by arrangement of their plates. Be careful not to slip while climbing on rocks to observe barnacles; the outer edges of the shell are sharp and can cause painful cuts to unwary or careless beachcombers.

Barnacles have a dual growth problem. As an arthropod with a chitinous exoskeleton, the skeleton must be periodically shed by molting to allow for body growth. Also, the calcareous volcano-shaped shell must be enlarged to accommodate growth. As new material is added to the outside, the inner surface is dissolved, probably by chemical secretion.

Barnacles are hermaphroditic (monoecious, with both male and female reproductive organs present in each individual) and fertilize neighbors' eggs by inserting a penis (a retractable tube) into them. The fertilized eggs hatch within the barnacle's shell and the larvae are released into the water. After a two-month free-swimming larval stage, they become permanently attached with a drop of brown cement. The cement is produced by glands in two sensory tentacles on the heads of the larvae. Then, they secrete the volcano-shaped calcareous shell around themselves, within which they spend the rest of their lives. The calcareous plates increase in size as the animals grow. They are often so numerous that the larval stage has difficulty finding a suitable spot for attachment.

Barnacles are filter feeders that have adapted to spending most of the time exposed to air. Their shell plates make them look like mollusks, but they have jointed appendages like other crustaceans such as lobsters, shrimps, and crabs. The fanlike appendages, called cirri, are extended out of the shell's opening to sweep through the water for feeding (Figure 3-7, and Color Plate 9d). They trap microscopic food particles in their hairlike bristles, then pass edible pieces to their mouth. Barnacles have been described as "animals that lie on their backs and spend their lives kicking food into their mouths with their feet." On a sunny day, if one closely observes water covered barnacles, one may see tiny shadows flickering everywhere over the submerged rocks, the fanning action of cirri.

Barnacles foul the bottoms of ships, adding as much as 200 tons to a large vessel and an increase in fuel consumption of 20% or more because of the drag. Ships periodically enter dry dock, for scraping off the attached organisms.

Goose neck barnacles (*Lepas* species) are also called stalked barnacles. They owe their name to their conspicuous, long, flexible, muscu-

Figure 3-7. Acorn barnacle with feeding appendages extended.

lar, necklike means of attachment and the shape of their shells (like the head of a goose). The calcareous shell plates are flattened and close the orifice without valves (Figure 3-8). The shell can be as much as two inches long, and the fleshy stalk may exceed four inches in length. Although they are found in the mid-intertidal area, they are more prevalent on floating objects such as driftwood. In Europe, goose neck barnacles are eaten raw or steamed.

Two other filter feeders that are often found in the crevices of intertidal zone rocks are the blue mussel and ribbed mussel (bivalve, two parts to the shell, mollusks). Mussels are the most abundant bivalves found on the rocks. However, ribbed mussels are more common in salt marshes. Food particles are trapped by mucus-covered gills and passed to the mouth by microscopic hairlike projections called cilia.

Figure 3-8. Goose neck barnacles (*Lepas* spp.) have a conspicuous long muscular neck-like means of attachment.

Mussels attach to a hard substrate or each other by strong fibers called byssal threads. The threads are secreted by a special gland in the foot. The foot's tip is pressed against a hard surface and the glandular secretion passes down a groove to the tip of the foot. A byssal thread is formed when the secretion is exposed to seawater and hardens. The foot is then moved to another spot, where the process is repeated until many "guy lines" hold the mussel firmly in place (Figure 3-9). Mussels are capable of locomotion, but they are very slow and have to reattach at the new location. The moist shelter of mussel beds may house other animals, such as worms.

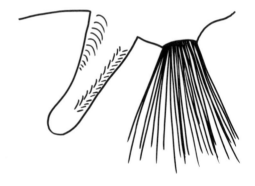

Figure 3-9. The mussel's foot is thin and extensile. The byssal gland in the foot secretes a thick fluid that runs along a groove and when exposed to sea water hardens into a tough thread. The gland is probably a modification of the gland used by snails to produce the mucus the foot glides over.

The delicious blue mussel (*Mytilus edulis*), also known as the common edible mussel, like oysters was a staple in coastal Native Americans' diet. Like many bivalves it is a filter feeder and the gills might possess disease-causing organisms if taken from polluted waters. The smooth, thin shell is long and narrow. The umbo is at the apex of the shell with prominent growth rings radiating from it. Exterior color is bluish black (occasionally brown rays are present); interior color is pearly white with a purplish blue margin. The blue mussel grows to about three inches long and is found from North Carolina to the Arctic (Figure 3-10, and Color Plates 7c and 7d).

The ribbed mussel (*Modiolus demissus*) has prominent radiating ribs on the shell that easily distinguish it from the blue mussel (Figure 3-11, and Color Plate 8a). Shell color ranges from blue-black to yellowish-brown or greenish-yellow. It has a bitter taste and is usually not eaten

Figure 3-10. The valves of a 3-inch-long blue mussel (*Mytilus edulis*).

Figure 3-11. The ribbed mussel (*Modiolus demissus*) has prominent radiating ribs on the shell that easily distinguish it from the blue mussel. The specimen is 3.5 inches long.

by humans. However, some say it is fuller-flavored than oysters. Regardless of the controversy over its edible quality, flounder like the ribbed mussel, and it is frequently used as bait. The bivalve is also found throughout the study area, but is more frequently seen in salt marshes and estuaries.

Microciona prolifera, commonly called the red beard sponge, can be found just beneath the low-water mark on hard surfaces. The color is bright red to orange-brown. It encrusts on rocks in the lower inter-tidal zone, but assumes a fingerlike, branching form, below the low-water mark. The scattered oscula are inconspicuous. It can survive in brackish water.

The calcareous tubes of polychaete worms, coiled in some species, are cemented onto rocks and seaweed. Featherlike radioles extend from the protective tube to filter food particles out of the water when the tide is in.

Hydroides dianthus is commonly called the carnation worm. Its long, twisted, not coiled, white calcareous (limestone) tubes are frequently seen on shells and rocks (Figure 3-12). A collar of tissue near

Figure 3-12. The calcareous tubes of the carnation worm (*Hydroides dianthus*) cover this rock, which is 6 inches across.

the head secretes the tube, which can be approximately three inches in length. The greenish tinted collar is translucent. The head bears a crown of radioles for food gathering and respiration. One radiole is modified into a pluglike operculum, which is used to close the tube opening when the radioles are withdrawn. If the tube plug is lost, a new one will be produced from a short filament on the other side of the head, which seems to be held in reserve just for this contingency. The radioles are purple, banded with white and green, or brown, banded with yellow and white. Blood vessels with green blood can easily be seen in the worm's almost translucent body. The sexes are separate and fertilization occurs in the water outside the tube. They are abundant on rocks and shells from Cape Cod southward.

Spirorbis species are frequently-encountered worms that build small, flat, coiled, snail-like, white, calcareous tubes. Thus, they are commonly called coiled worms. The coiled tubes of *S. borealis* are about ⅛-inch in diameter with left-handed coils, and usually attach to *Fucus* (rockweed) and other seaweeds attached to the rocks, and occasionally directly to the rock. *S. violaceus* has right-handed coils and three ridges running its entire length; it is usually found on rocks and shells, seldom on seaweeds. *S. spirillum* also has right-handed coils, but a smooth tube; it is found on a variety of substrates. *S. borealis* and *S. spirillum* are usually found in the intertidal zone while *S. violaceus* is found in deep water. All possess delicate plume-like feeding radioles like *Hydroides*, but the operculum is funnel-shaped. When the tubes are exposed at low tide, the operculum closes the opening to prevent desiccation. They are hermaphrodites, and fertilization and development of the embryo occur within the tube. *Spirorbis* species are abundant from Long Island northward (Figure 3-13).

Sea anemones may be attached to rocks in the lower area of the intertidal zone, in tidal pools, or just beneath the low water mark. These large, cylindrical polyps are often quite colorful; they are usually found attached by their adhesive pedal disk to rocks, wood, shells, etc. Most anemones can move about if necessary, inching along on the muscular pedal disk. The disk is expanded in one direction and then contracted, pulling the body into the new position. Care should be shown in collecting these animals because it is easy to tear the body wall trying to remove them. They are solitary creatures and do not form colonies.

Figure 3-13. The coiled, snail-like, white tubes of *Spirorbis* sp. on the brown alga *Fucus*.

Sea anemones, sedentary relatives of the roaming jellyfish, have a flowerlike appearance when covered with water with their tentacles extended. An expanded anemone has grace and beauty. The tentacles possess stinging cells to catch prey, but a finger inserted into the tentacles of a sea anemone in this study area produces only a tingling sensation or stickiness.

The most common sea anemone in the northern region of the study area is the frilled anemone (*Metridium senile* [*dianthus*]), also called the common sea anemone or the northern anemone. The large anemone's column may exceed four inches in length and three inches in width. Masses of short tentacles are present, forming a feathery mass. The tentacles are unsatisfactory for the capture of large organisms, but are proficient at capturing plankton, its primary food. Hairlike cilia move the plankton to the mouth, located in the center of the mass of tentacles. Offshore rocks and shipwrecks (artificial reefs) are literally covered with the large, colorful anemones that range in color from white to dark brown—even pink (Color Plates 4b, 4c, and 4d). In shallow water small specimens are usually found in dark places, such as among seaweed and the undersides of rocks. With tentacles and column contracted, they look like small volcanoes. They are common as far south as northern New Jersey.

Sea stars (starfish) also frequent the intertidal area. These slow-moving echinoderms are carnivores; they use their tube feet to open mussels, scallops, and other shellfish. They also manage to feed on the spine-covered sea urchin, another echinoderm. The common sea star is the eastern starfish (*Asterias forbesi*) (Color Plates 2b and 13b). It usually has five arms—but variations are common—and four rows of tube feet on the undersurface of each arm. The tube feet are part of the water vascular system, described in Chapter 1. Pick up a sea star, turn it over, and observe the tube feet waving about in search of a solid substrate to attach to. The mouth is at the center of the star. Turn the sea star onto the other side and look with a hand lens for dermal branchia, small fingerlike respiratory structures. Note the small madreporite, a small, orange, circular structure that filters water as it enters the water vascular system.

Carnivorous snails roam the intertidal zone, feeding on barnacles and mussels. Some snails use their muscular foot to force open the shell of a barnacle, then the radula rasps away the flesh. In some species a narcotic secretion called purpurin is secreted to reduce the resistance of a victim.

Some snails, such as the dogwinkle (*Thais* [*Nucella*] *lapillus*), use their radulas to drill a hole through the shells of mussels to get at the soft body parts inside (Figure 3-14). Its foot secretes an acid to weaken the prey's shell. It also feeds on limpets and periwinkles, but prefers barnacles and mussels. An adult has a thick white shell with a thickened lip and about five whorls. The rounded spiral ridges give the shell a corrugated appearance. Color variations can be yellow, orange, brown, or banded with white, as determined by the snail's diet. The shells of snails that feed primarily on barnacles are white, while the shells of snails feeding mostly on mussels are dark. Shell size is up to 1½ inches (Figure 3-15). Anal glands secrete purpurin, which was used by ancient Phoenicians of the Mediterranean coast to produce the famous Tyrian purple dye and by monks to illuminate their manuscripts. Native

Figure 3-14. Diagram of a snail head, showing the rasping tongue-like radula (A). Photograph of a radula (magnified 40 times) showing dark-colored teeth (B).

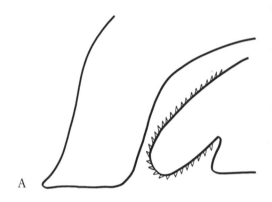

A

B

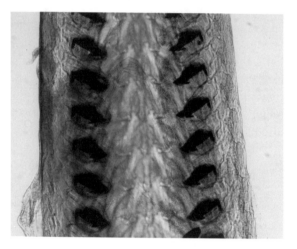

Americans also used the dye. The snail is common from Long Island northward, but less abundant on exposed rocky shores. It is also found in bays and estuaries. The opening of a dogwinkle shell has a deep groove through which the siphon extends. The opening of a periwinkle is smooth, lacking this groove. A siphon to pull water in for respiration is characteristic of snails living on mud or sand bottoms rather than for those of hard substrates. A siphon extending above mud pulls in clean water. The ancestors of the dogwinkle probably moved from being soft bottom dwellers to the rocky shore.

Dogwinkles are often confused with waved whelks (*Buccinum undatum*). The aperatures of the waved whelk are usually larger and the shell

Figure 3-15. The dogwinkle (*Thais lapillus*) has a short spire and thick outer lip. Often longitudinal, internal folds are present on the outer lip. The specimen on the left is 1 inch long.

can be four inches long. Also, the body is white with black spots (Color Plate 7a).

Sessile (non-motile) mussels are not totally helpless against the attacks of shell-boring snails. Some take advantage of the fact that the predator will attack only live organisms. They often respond by appearing to be dead, withdrawing as far as possible within a shell that is gaping open slightly, as though dead. The ruse frequently works. Another effective defensive measure is to tie an attacking snail down with byssal threads, rendering it helpless. Dead snails, tied down in that manner, are occasionally found in mussel beds.

Rocks in the lower region of the intertidal zone are often encrusted with filter-feeding sponges. Sea slugs (nudibranchs) are shell-less gastropod mollusks, like garden slugs. They are graceful and beautiful, contrary to their name, and prey on sponges, hydroids, and anemones. Sea slugs lose their shell while still microscopic in size.

Aeolidia papillosa is commonly called the plumed sea slug. Its back is covered with hundreds of club-shaped cerata, external gills, in oblique rows (Color Plate 5b). The name nudibranch means naked gills. Two pairs of tentacles are on the head. Color varies from gray to orange. This nudibranch reaches a length of approximately four inches. It is carnivorous, feeding on hydroids and sea anemones, and can use its prey's undischarged nematocysts by transferring them to cnidosacs (that open to the exterior) at the tips of its own cerata, where they are used by the nudibranch for defense. It is found in tide pools, the intertidal zone, and at great depths from Long Island northward.

Coryphella rufibranchialis is a beautiful nudibranch commonly known as red-fingered sea slug. It can easily be distinguished from the plumed sea slug by its smaller (usually less than an inch), more slender white body and cerata that occur in six or seven clusters. The cerata are red with white tips (Color Plate 5c). The nudibranch feeds on sponges, hydroids, and other invertebrates. It is found on the bayside of Cape Cod northward, seldom on the south shore of Cape Cod.

Various species of crabs also inhabit the area, scampering from crevice to crevice and in and out of tide pools. Many feed on seaweeds at night. Shore crabs must return to the water periodically to wet their gills.

The rock crab (*Cancer irroratus*) is usually subtidal in the study area, but occasionally will be found in crevices on the jetties or under rocks, and dead crabs and casts wash ashore. The upper surface of the carapace is dark yellow with small reddish-purple spots and is covered with fine granulations. The margin of the carapace has nine broad teeth. The crab grows to over five inches.

A crab similar in appearance and easily confused with the rock crab is the northern crab or Jonah crab (*Cancer borealis*). The margin of the carapace is not as smooth as that of the rock crab and the upper surface of the carapace is brick-red in color with purplish spots. The Jonah crab is found throughout the study area, but is more common in New England, especially north of Cape Cod. It grows to six inches and is found in tidal pools and under rocks (Figure 3-16).

Carcinus maenas, the green crab, described in the salt marsh habitat, is found under rocks in the intertidal zone, in tide pools and in crevices on jetties. It is easily distinguished from the rock crab and Jonah crab

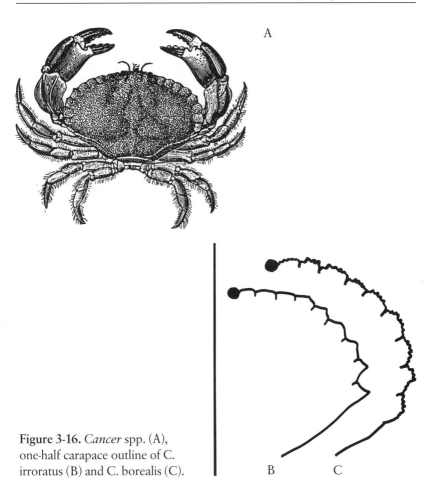

Figure 3-16. *Cancer* spp. (A), one-half carapace outline of C. irroratus (B) and C. borealis (C).

by the green color of the upper shell, and is generally not more than three inches long (Color Plate 13a).

The supralittoral splash zone lies above the intertidal zone. Its marine organisms are limited by the level to which salt spray lands. It is the habitat of lichens, a combination of algae and fungi, and some snails and crabs. A small snail called the common periwinkle (*Littorina littorea*), also called the edible periwinkle and the European periwinkle, is the dominant animal in this area. It feeds on seaweeds in the intertidal zone. The shell is a thick cone with whorls that average about an inch in height. Shell color can be yellow, gray, olive, brown, or almost black; it is often banded with brown or red. The interior surface is white with a brownish-black lip (Figure 6-11). The periwinkle's thick, solid shells are able to withstand the force of waves breaking on the rocks. It does not hold on with the same tenacity as do limpets. It finds safety from the full force of the waves by retreating into crevices of the rocks or among seaweed. The common periwinkle when exposed by the tide will often secure itself to the rock by mucus. When the mucus hardens, the foot is withdrawn into the shell and the operculum closes the opening. It is a precarious position for the snail, even a hard wind can topple it from its perch. Tons of periwinkles are sold as food in European markets; they are even roasted in their shells and sold from street corners. They extended their range to the United States during the 19th century, probably on the hull of a ship, and are now found as far south as Maryland.

The southern periwinkle (*Littorina irrorata*), also known as the Gulf periwinkle, is common from Maryland southward. The shell is more pointed than that of *L. littorea* and has spiral rows of dark spots. *L. irrorata* are more frequently found in the salt marsh than on the rock jetties.

The rock lice and the sea roach are the unflattering common names of isopods that belong to the genus *Ligia*. These flattened crustaceans have several pairs of legs and may exceed an inch in length (Figure 3-17).

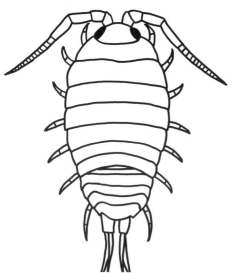

Figure 3-17. Sea roaches (*Ligia oceanica*) are large, fast-moving isopod crustaceans that hide in rock crevices during the day.

They hide in rock crevices around the high tide line, moving about at night when the tide is out and feed on organic debris, such as decaying seaweeds carried in by the waves. They are nocturnal to escape predation by shore birds. Sea roaches scurry out of a flashlight beam and only a few would be out in bright moonlight. They venture into the water only briefly, and will drown if submerged for long.

PLANTS OF THE ROCKY REALMS

As mentioned earlier in the chapter, seaweeds usually dominate the rocky habitat. Whenever there is an outcrop of rock or stones, there will be a rich and varied community of seaweeds in the intertidal zone. Their broad fronds provide attachment for hydroids, bryozoans, and other animals. They also are a source of food and a protective blanket against predators and wave action. Many animals find shelter from exposure to air, excessive heat of summer, and frigid cold of winter within the dense growth of seaweeds. Rocks are covered with seaweed at all seasons. There is constant loss and renewal of plants throughout the year. Minute attached sporelings are always present to replace the older plants torn away by surging tides and storms.

The most conspicuous plants in the intertidal zone are brown seaweeds, members of the genus *Fucus*, commonly called rockweed. Their thick cell walls reduce water loss while the plant is exposed to the air at low tide. Lift up the mats of rockweed to search for small animals.

The lower region of this intertidal zone is submerged most of the time; it is exposed for only a couple of hours on a few days each month, at spring tides. Large brown seaweeds known as kelps are most abundant, but there are also red seaweeds, such as Irish moss (*Chondrus*) and the delicate, paper-thin laver (*Porphyra*), which are collected for food. Calcareous red algae develop as tufts on some rocks and as an encrustation on others.

Brown and red seaweed can live lower in the intertidal zone, and underwater at all times because of their pigmentation. The brown pigment fucoxanthin and the red pigment phycoerythrin absorb the light wavelengths of the underwater environment that cannot be absorbed by chlorophyll.

Dense seaweed growth provides protection for many marine animals. The plant mass protects them from drying as the tide recedes. Some use the seaweeds as a substrate to live on. Seamats (bryozoans) frequently develop large colonies on seaweeds and rocks. The tiny animals are also referred to as moss-animals because they live in encrusting colonies, and are often mistaken for plants. Small tube worms (annelids) and hydrozoans (cnidarians) are frequently found living on seaweeds. An isopod found on seaweed and in tidal pools along rocky shores is *Idotea baltica*. The crustacean is about one inch long and is often green, but color is variable (Figure 3-18).

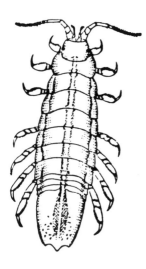

Figure 3-18. *Idotea baltica,* an isopod crustacean.

Green Algae

Bryopsis plumosa, one of the most beautiful algae, is a filamentous form that has branches with branchlets extending from the main axis. This type of branching often creates a triangular, featherlike appearance, and gives rise to the common name sea fern.

Bryopsis occurs on rocks around the low tide mark and sublittoral. It grows to four inches, but is not easily seen because it often grows under ledges on the rocks (Figure 3-19).

Figure 3-19. A 4-inch-long specimen of the sea fern, *Bryopsis plumosa*.

Brown Algae

These algae are much more numerous. Those in the intertidal zone attach by a disclike holdfast. The following are examples of brown algae, but in addition to the brown pigment fucoxanthin, they also contain the green pigment chlorophyll. Usually, the brown pigment masks the green, but not always. Some of the brown algae listed here may appear green, and can cause confusion for the less experienced beachcomber.

Desmarestia viridis is a filamentous alga with opposite branching. Unlike most species of this genus, *viridis* has cylindrical, rather than flattened filaments. Plants grow to over six inches, and are found from New Jersey northward, low in the intertidal zone (Figure 3-20).

Petalonia fascia has thin, broad blades, but lacks a conspicuous stipe or holdfast. Cells of different sizes are apparent when the thallus is cut in cross section and observed with a microscope. They range in size to

Figure 3-20. A 6-inch-long specimen of the filamentous brown alga *Desmarestia.*

about ten inches long and one inch wide. *Petalonia* can be distinguished from a young *Laminaria* because of the well-developed holdfast of the latter (Figure 3-21).

Punctaria species are similar in appearance to *Petalonia*, but they are more slender and delicate, giving rise to the common name ribbon weed. Cells of equal size are observed when the thallus is cut in cross section and observed with a microscope. They are often found attached to other seaweeds and eel grass.

Scytosiphon lomentaria is a long hollow tube tapered at both ends. Constrictions occur along the length of the tube, making it easily identifiable and is the reason it is often referred to as sausage weed (Figure 3-22). It is found low in the intertidal zone on jetties exposed to rough water and in the quiet water of bays. Plants usually grow in clumps, giving the appearance that *Scytosiphon* is one large plant branching at

Figure 3-21. The brown alga *Petalonia fascia* does not have a well-developed holdfast like that of *Laminaria*. The specimens are about 6 inches long.

Figure 3-22. The constrictions along the length of *Scytosiphon lomentaria* give the brown alga the common name sausage weed. The blade on the right is 13 inches long.

the base into several tubes. However, each plant is one unbranched tube that can grow to almost two feet long and one half inch wide.

Devil's bootlace (*Chorda filum*), also known as cord weed, has a solid, unbranched, cordlike frond. Young plants are usually covered with hairs (Figure 3-23). The terminal parts of old plants are usually frayed. The long, unbranched frond can be a quarter of an inch in diameter and up to 16 feet in length. The surface is slimy and the fronds are exceptionally tough and hard to break. The plant is found more frequently on shallow sandy bottoms. It is usually found near the low water mark from Long Island Sound northward.

All large brown algae such as *Laminaria* species are commonly called kelp. They are the largest algae found in the study area, and are found from Long Island northward (Color Plates 2b and 14b). *Lami-*

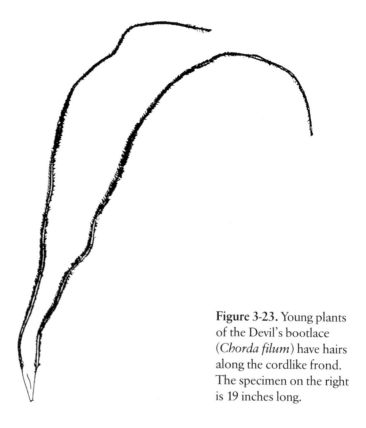

Figure 3-23. Young plants of the Devil's bootlace (*Chorda filum*) have hairs along the cordlike frond. The specimen on the right is 19 inches long.

naria are most often found below the low water mark. They grow in turbulent water close to shore, and are often washed up on the beach after severe storms. The three basic body parts (blade, stipe, and holdfast) are conspicuous. The leatherlike blade does not have perforations or midribs. Growth does not occur at the tip as in land plants, but from a point just above the stipe. A new blade is produced annually, pushing away the old one. If growth occurred at the tip, the pounding surf would strip away the cells producing the new growth. *Laminaria* species have large branching holdfasts, resembling a mass of roots, anchoring them to the substrate. These plants are easily recognized by their large size.

Laminaria agardhii may exceed ten feet in length and eight inches in width. The leatherlike blade is broad, flat, and often frayed at the end. The blade narrows into a large cylindrical stipe, which varies in length from less than an inch to more than one foot (Figure 3-24).

Figure 3-24. A young 9-inch-long specimen of *Laminaria agardhii.*

Laminaria digitata is often referred to as horsetail kelp or devil's apron. It has a large, deeply-divided, fan-shaped blade, and grows to about four feet (Figure 1-43).

Agarum cribosum has a prominent blade, stipe, and holdfast. The alga is easily identified by its perforated and undulated blade, and is commonly known as sea colander. *Agarum* is found from Cape Cod north (Figure 3-25).

Fucus species are one of the most predominate marine algae in the study area. The color is green-brown, and the holdfast is disclike. The flattened, dichotomous (forking) branches have prominent midribs. The strengthening midrib enables the plants to better withstand the pounding waves of exposed areas. When sexually mature, the swollen tips of many of the branches are covered with small, light colored spots (conceptacles) in which the gametes (sex cells) are produced. *Fucus*

Figure 3-25. The perforated blade of *Agarum cribosum* denotes the common name sea colander. The young specimen is 5 inches long.

species are called rockweed because they literally cover the rocks of the intertidal zone in the northern part of the study area.

Fucus vesiculosus has paired vesicles (air bladders) along the midrib, and the branches are straight (Figure 3-26). These swellings contain oxygen and other gases produced during photosynthesis. A common name for the plant is bladder wrack. It is often called popweed by children because of the sound produced when its vesicles are compressed and broken between fingers. During the reproductive season, the tips of the branches swell into bulbous, somewhat heart-shaped structures that produce the reproductive cells. It is found from North Carolina northward, in the middle of the intertidal zone.

Figure 3-26. The rockweed *Fucus vesiculosus* can be identified by the paired vesicles along the midrib.

Fucus spiralis does not possess vesicles, and its branches are spirally twisted but seldom make a complete turn (Figure 3-27). It is often referred to as spiral wrack. The midrib extends up the receptacle, an enlarged reproductive structure present at the tip of the branch during some seasons, forming a ridge. It is found from Long Island northward.

Ascophyllum nodosum has main branches that are flattened, with large single vesicles and no midribs. The air bladders are large, single, and cover the full width of the frond. Small club-shaped reproductive branchlets extend off the main branch (Figure 3-28). *Ascophyllum* is sometimes called knotted wrack and is quite variable in form and size. The plant is almost black in color; it can be distinguished from *Fucus* by the absence of a midrib. It is found from Long Island Sound northward, in the middle of the intertidal zone, in sheltered waters.

Figure 3-27. The rockweed *Fucus spiralis* does not possess vesicles. The specimen without its holdfast is 6 inches long.

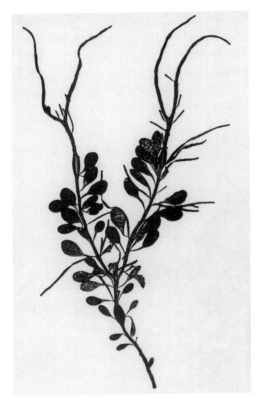

Figure 3-28. *Ascophyllum nodosum* has large single vesicles that cover the full width of the frond. The specimen, missing the lower portion, is 12 inches long.

The air bladders of *Fucus* and *Ascophyllum* buoy the plants when submerged, allowing better light penetration for photosynthesis. The cell walls of these plants contain gel-like compounds that protect them from desiccation when exposed at low tide and provides some protection from pounding waves.

Red Algae

The red pigment, phycoerythrin, of these algae usually masks the green pigment chlorophyll. However, like the brown algae, color is sometimes an unreliable means of classification.

Porphyra species are a membranous form, similar in appearance, except for color, to the green algae *Ulva* and *Monostroma* described in the bay habitat. *Porphyra* is an almost transparent plant, with a light pink-brown to purple-red color. The genus name *Porphyra* means "a purple dye." The plant body is a single sheet of cells and the flattened plants are identified by variations in the outline of the frond (Figure 3-29). The plant, commonly known as laver, is attached by a small cord of interwoven strands; hence, its specific name *umbilicalus*.

Figure 3-29. The membranous red alga *Porphyra* is a major food staple in the Orient. The specimen is 7 inches in length.

Porphyra is a major food staple in the Orient. In Japan, it is grown commercially by stringing nets between pilings and seeding them with chopped *Porphyra*. Within a year, the plants are harvested and ground to a paste, then pressed and dried into a solid mat about eight inches square. It is often wrapped around cooked rice in Japanese restaurants and sushi bars.

Grinnellia americana is a membranous form similar in appearance to *Porphyra*. Unlike *Porphyra*, it has a prominent midrib. *Grinnellia* is one cell thick near the margins, thicker near the midrib. The characteristic dark spots on the thallus are reproductive structures.

Corallina species have an erect, flattened, pinnately-branched (branches on opposite side of the main axis, as in a feather) thallus growing to more than four inches. They are unique in the study area in that they are the only erect calcareous algae. The calcareous algae are members of the Corallinaceae family, so named because of their superficial resemblance to coral. Until the mid 1800s they were thought to be corals. The calcium carbonate is deposited in bands, producing flexible joints, which allow the alga to bend with waves that would otherwise shatter the delicate plant. These calcareous segments make *Corallina* species, sometimes called pink coral, easy to identify (Figure 3-30); they are responsible for the pink-to-white color on many rocks. Their calcareous covering may be to protect the alga from plant-eating animals.

Phymatolithon laevigatum is another calcareous alga but it is not erect; it forms a pink or white crust on rocks (Color Plates 2d and 6c). The thallus is several cells thick. Sea urchins feed on the algae to get calcium carbonate to maintain their own calcareous skeletons. The plant is sublittoral but can be seen just beneath the surface in tide pools.

Hildenbrandia ruba is not a calcareous alga, but is similar in appearance to *Phymatolithon*. The thallus is thinner, and the plant is usually red in color—not white as is often the case with *Phymatolithon*.

Figure 3-30. Pink coral (*Corallina*) is the only erect calcareous alga in the study area. The specimen is about 1 inch long.

Chondrus crispus, commonly called Irish moss, varies greatly in color from yellow-green to an almost iridescent purple-red. Blades are flattened, dichotomously branched, narrowing downward to the point of attachment, and are often curled (Figure 3-31). During the reproductive season, dark spots on the blades are where reproductive cells are produced. The plant grows in a shrublike form up to four inches high. It is commonly found on rocks in the lower intertidal zone and below, and is abundant from Long Island Sound northward.

For decades, Irish moss was imported from Britain to Boston by the Colonists to make their famous blancmange custards; then it was discovered that it grew in abundance along the New England coast. Cooked with milk, seasoned with vanilla or fruit, *Chondrus* makes a

Figure 3-31. The red alga Irish moss (*Chondrus crispus*) was used by Colonists to make their famous blancmange custards. The specimen is 3.5 inches long.

highly palatable custard. The alga is an important commercial species in northern New England. *Chondrus* has been harvested commercially for more than a century, and is the source of a phycocolloid called carrageenin. The compound has many uses as a stabilizer; for example, it keeps chocolate suspended in commercial chocolate milk. It is also used in making jellies, candy, and salad dressing. A handful of Irish moss cut up and cooked with soup or stew provides a nourishing and, unlike flour, non-fattening thickening. Chinese restaurants frequently use *Chondrus* as a thickener in hot and sour soup.

Rhodymenia palmata is commonly referred to as dulse and, like *Chondrus*, varies in appearance. The short stipe broadens gradually into flattened blades, which are dichotomous or palmately branched (branches arise from a common point, like fingers from a palm) (Figure 3-32). They are deep red-purple in color. Small leaflets often grow from the margin of their blades. Plants are usually smaller, but some more than 12 inches long have been found in the intertidal zone and below. It is

Figure 3-32. Wads of dried dulse (*Rhodymenia palmata*) are often chewed like tobacco. The specimen is about 8 inches long.

abundant from Long Island northward. This alga is also collected commercially for use as a thickener for soups and sauces. In Nova Scotia and some areas of Europe, wads of dried dulse are chewed like tobacco.

Ceramium species are common filamentous algae that can be easily identified by the alternating light and dark horizontal bands on their main axis and branches. These bands, visible without magnification, are due to layers of cells that cover the thallus, masking underlying tissue. The filaments are dichotomously branched, with pincerlike curved tips, giving rise to the common name pincher weed. Color varies from pink to brown-red.

Callithamnion corymbosum, with its irregular to alternate uniseriate branching, produces very delicate spherical masses that have the appearance of pink clouds in the water. The plants grow to four inches in diameter below the low water mark (Figure 3-33). By late summer they often lose their obvious red color, becoming brown-red. They are usually epiphytic (attached to) on other plants. An epiphyte lives on another plant but is not parasitic.

Polysiphonia species are filamentous with many branches, producing stiff tufts. Color ranges from purple-red to almost black. They can be found as epiphytes on other seaweeds, or attached to shells and rocks (Figure 3-34). Size ranges from a few millimeters to 12 inches. Most *Polysiphonia* species are seldom found in the intertidal zone; like most red algae they prefer to be covered by water at all times. Plants of

Figure 3-33. *Callithamnion corymbosum* often forms pink spherical masses in the water. The specimen on the left is 1.5 inches long.

Figure 3-34. A 1.75-inch-long specimen of *Polysiphonia elongata* growing as an epiphyte on eel grass (*Zostera marina*).

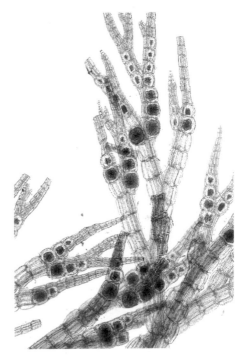

Figure 3-35. *Polysiphonia* sp. magnified 40 times. The dark areas are tetraspores, asexual reproductive structures.

Figure 3-36. The branchlets of *Rhodomela subfusca* end in tufts. The specimen, growing as an epiphyte on rockweed, is 4.5 inches long.

this genus may be readily recognized in the field, but identification of species requires magnification, to count the number of pericentral cells in the branches (Figure 3-35).

Rhodomela subfusca has cylindrical axes with small branchlets ending in tufts that may be lost late in the season (Figure 3-36). When the tufts are absent, it is often confused with *Polysiphonia*.

JETTIES ARE WORTH VISITING

Large rocks such as those found on jetties (groins) offer an excellent substrate for animals and plants to attach to. However, the vast array of marine organisms are only fully revealed to those who search diligently. Look in tide pools, narrow crevices in rocks, under rocks, among the intertidal seaweeds, and even the large holdfasts of kelps. Rocks that are turned over should be replaced in their former position.

Large rocks are a far more productive area for observations of marine organisms than a sand beach. However, be careful scrambling over rocks in search of marine organisms; barnacles will cut, and wet rocks and algae are slippery; wear rubber soled shoes. Remember to watch for waves.

4

Pebble Beaches and Man-Made Substrates

Even a blind man could not stand upon a shingly beach without knowing that the sea was busily at work. Every wave that rolls in from the open ocean hurls the pebbles up the slope of the beach, and then as soon as the wave has broken and the water has dispersed, these pebbles come rattling down with the currents that sweep back to the sea.

Thomas Henry Huxley

PEBBLE BEACHES

A pebble or gravel beach cannot support much animal or plant life, nor can many survive in the rattling mill of the pebbles. The wrack line at the high-water mark is the most promising area to search for marine organisms.

Many of the stones in the intertidal zone have pretty color patterns. You might want to collect some to take home; they can be used as paperweights or knickknacks. Unfortunately, when the stones dry they usually lose their luster. To return their lost brilliance, scrub them with soap and fresh water. After they dry, take them outside or into a well ventilated room and spray them with several coats of clear acrylic plastic. After they dry, turn them over and coat the other side; they should then be as shiny as when they were first seen at the beach. The acrylic plastic takes the place of the water and fills in the tiny surface scratches that were responsible for their dull appearance.

Colorful stones can be made into an attractive mosaic when glued to a flat piece of driftwood. A combination of stones and shells is often used in creating a piece of seashore art. Pieces of glass recovered from a sand beach, frosted by tumbling in the surf and sand, may also be incorporated into the mosaic. Sand-polished pieces of glass bottles can look attractive if placed in a decanter, covered with water, and placed on a windowsill where the sun may shine through them. Seashore art can take many forms, such as the molted shells (never sacrifice living animals for art) of young horseshoe crabs or other animals on a piece of driftwood (Figure 4-1).

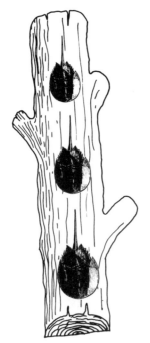

Figure 4-1. The casts (molted shells) of young horseshoe crabs glued to a piece of driftwood make a nice example of seashore art. Always use casts, not living animals.

Pebble beaches may not have a proliferation of organisms, but large rocks such as those found on jetties (groins) offer an excellent substrate for animal and plant attachment.

MAN-MADE SUBSTRATES

Wood, concrete, and iron pilings, wood planks of docks and break-waters, and boat bottoms also offer excellent substrates for animals and plants known as the fouling community. The point of attachment, like on a rocky coast, depends upon the organism's ability to withstand air exposure and temperature changes during the receding tide. The animals feed on water-borne detritus, plankton, or each other. The first animals to settle on the man-made substrates are barnacles; then sponges, hydroids, oysters, bryozoans, and other sessile forms follow. These in turn provide a vast habitat for a multitude of small organisms including amphipods, isopods, the bizarre skeleton shrimp, nudibranchs, tube-dwelling worms, small crabs, anemones, and tunicates. The fouling community is a rich source of food that attracts large numbers of fish.

Most organisms found on rocks will also attach here, with a greater abundance of the delicate hydroids and erect forms of sponges if the pilings are in quiet water.

Two hydrozoan cnidarians, *Obelia* and *Tubularia*, are found attached to hard surfaces. *Obelia* species have polyps and medusae body types in their life cycle. When in the polyp stage, they develop into a colony two to eight inches in height, with two types of polyps. One is tentacled and specialized for feeding (nutritive polyp). The other lacks tentacles, and serves for reproduction (Figure 1-7). The different types of polyp make it easy to identify *Obelia* with a hand lens (Figure 4-2). The colony has a feathery appearance because of its many white branches (Color Plate 2b). The branches bear polyps, are hollow, and contain a gastrovascular cavity that is continuous with the cavities of the polyps, so nutrients can be distributed throughout the colony. The polyps of colonial hydrozoans are connected to the branches at the base of the column. The branches of *Obelia* and many other colonial hydrozoans are covered, and are protected by a transparent chitinous (a complex derivative of carbohydrates) structure called a perisarc, which is secret-ed by the epidermis. The perisarc, which continues up the polyps, is

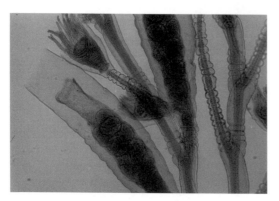

Figure 4-2. *Obelia* sp. feeding polyp (top left, with tentacles) is easily distinguished from a reproductive polyp with developing medusae (below left). Note the protective, transparent perisarc. Magnified 40 times.

called a hydrotheca on the nutritive (hydrocanth) polyp and a gonotheca on the reproductive (gonangium) polyp. The medusa stage is small and inconspicuous (Figure 4-3). *Obelia* attaches to stones, pilings, eel grass, large marine algae like *Laminaria*, and other objects (Figure 4-4, and Color Plate 2b).

Tubularia crocea is pinkish-red in color and is commonly known as pink-hearted hydroid. The colony, somewhat smaller than *Obelia*, grows to approximately six inches in flower-like clusters. It is easily distinguished from *Obelia* in its color and because the colony is com-

Figure 4-3. *Obelia* medusae are small and inconspicuous. Magnified 40 times.

Figure 4-4. A colony of *Obelia*, attached to an alga, has a feathery appearance. (Photo by Norman Despres.)

posed of only nutritive polyps that lack a hydrotheca (Color Plates 2d, 3c, and 3d). Also, *Tubularia* has two whorls of tentacles; *Obelia* has only one. The nutritive polyp also produces the small medusae, which attach at the base of the long tentacles. Both hydroids are also found attached to various substrates throughout the study area.

Some pilings, rocks, and seaweed have areas encrusted with colonies of moss animals (bryozoans). Sea lace (*Membranipora* species) forms thin, white encrustations, where each individual animal (zooid) is clearly visible, producing a lace-like appearance (Figure 4-5, and Color Plate 2c). A hand lens will reveal small openings through which the animals extend their cilia-covered tentacles for filter feeding. Cilia are hair-like projections from cells. The delicately curved margins of the colonies are areas of active growth. Sea lace is often found on seaweeds such as *Ascophyllum* and *Chondrus*. Another bryozoan, the tufted bryozoan (*Bugula* species), grows into an erect, branching, treelike colony, about three inches in height, yellow to dark brown in color (Figure 4-6). Because of its erect growth, another common name for the animal is erect bryozoan, and it is often mistaken for a plant. However, the colony feels coarse or hard, not flaccid, because it is lightly calcified.

Figure 4-5. A green sea urchin (*Strongylocentrotus droebachiensis*) on the bryozoan sea lace (*Membranipora* sp.). Note the sea urchin's tube feet. (Photo by Jim Matulis.)

Figure 4-6. The tufted bryozoan (*Bugula* sp.) is an erect colony rather than an encrusting form like sea lace (*Membranipora* sp.). Magnified 40 times.

A small crustacean frequently found among bushy hydroids, bryozoans, and seaweeds is the skeleton shrimp (*Caprella* species). Actually, it is not a shrimp; it is an amphipod (like a beach hopper) not a decapod, and looks more like a praying mantis than a shrimp. It has a thin body with few appendages and grows to little more than one-half-inch in length. Like the praying mantis, it is a voracious predator that uses a pair of enlarged mantislike appendages to grab its prey, other small crustaceans (Figure 4-7, and Color Plate 10b). The skeleton shrimp itself is preyed upon by small fish like the sea horse.

Attached to pilings, and other hard substrates, especially in quiet water, are several tunicates. The sea grape (*Molgula* species) is grapelike in appearance and size, and is grayish in color, with conspicuous siphons that reveal its identity. *M. manhattensis* is the most common species; it can be found even in the polluted waters of New York Harbor (Figure 4-8). It often occurs in dense clusters. If you squeeze the

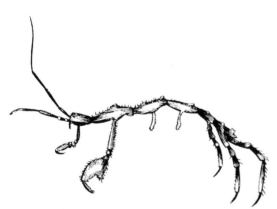

Figure 4-7. The skeleton shrimp (*Caprella* sp.), an amphipod crustacean, looks more like a praying mantis than a shrimp.

Figure 4-8. The sea grape (*Mogula manhattensis*).

sea grape, water squirts out of the siphons. Some tunicates are called sea squirts because of this behavior.

A common encrusting, colonial tunicate, *Botryllus schlosseri*, often referred to as the star tunicate or eyed tunicate, is found on pilings, eel grass and many other hard substrates. The flat, gelatinous colonies are often gold, but color is variable; it grows to one-eighth of an inch thick and about four inches in diameter. The embedded zooids (individuals of the colony) are arranged in unmistakable star-shaped clusters that are often white, yellow, brown, or purple in color (Color Plate 15a). Each animal draws in water to filter food and for respiration, but all members of a "star" share a common opening for its extrusion.

We normally associate corals with massive reefs, such as those found in the warm waters of the Caribbean Sea or off Australia. Many people are surprised to learn that coral lives in northeast waters. However, there is only one species, *Astrangia danae*, and it does not form reefs. The common name is northern coral or star coral. It belongs to a group referred to as stony corals because they produce an external calcium carbonate skeleton. These crystals are secreted by the epidermis of the lower half of the column, forming a cup-like structure (theca), into which the polyp can contract for protection. The column's septa also secretes calcium carbonate, producing thin, radiating skeletal plates that extend up from the cup's bottom. They are easily seen when observing a piece of coral skeleton without living tissue (Figure 4-9).

Figure 4-9. The external skeleton of northern coral (*Astrangia danae*). Note the cup-like thecas with radiating skeletal plates on this 2.5-inch specimen.

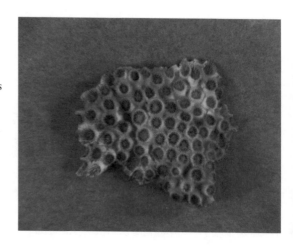

Northern coral does not form reefs, but it does produce large colonies of thin, encrusting masses with a starlike pattern. The polyps of colonial corals have lateral inter-connections that form a connecting sheet of tissue composed of upper and lower layers of gastrodermis and epidermis, with a gastrovascular cavity between. The extensions of the gastrovascular cavities allow food to be dispersed between polyps. The lower epidermis of the connecting sheets secretes skeletal material between the cups.

Northern coral is common from Cape Cod southward when a suitable substrate is available. The encrusting coral is usually found on pilings, rocks, or other hard surfaces at the low-water mark or below. Occasionally, erect branches are produced. The delicate little polyps, not more than ¼-inch in height, have 18 to 24 white-tipped tentacles. The polyps are well separated, giving a pitted appearance to the skeleton. The polyps are variable in color, but usually are white or pink, although occasionally they may be green or brown from the presence of algae in their tissues.

BORING ANIMALS

One boring animal found in this habitat, but not on rocks is the shipworm (*Teredo navalis*). Contrary to its name, the shipworm is not a worm but a bivalve mollusk that has become adapted for burrowing in wood. The shell is reduced in size, but the two small valves at the anterior end are modified to form rasping structures used for drilling cylindrical tunnels in wood. Thus, most of the long worm-like body, up to two feet in length and one-quarter-inch in diameter, is not enclosed by the valves (Figure 4-10).

A shipworm larva that comes in contact with a piece of wood will form a temporary attachment with a single byssal thread like that of a mussel. The tiny mollusk bores a hole less than one millimeter in diameter into the wood. The tunnel follows the grain of the wood, the course of least resistance, and is coated with a calcareous lining secreted by the mantle. It is confined within its tunnel, but the posterior end of the animal forms siphons that keep water flowing over the gills of the filter feeder. At the posterior end of the body, near the siphons, a pair of small limy plates are attached. These pallets are used to seal the minute

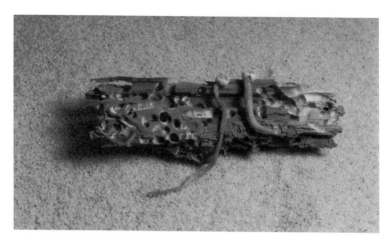

Figure 4-10. Two shipworms (*Teredo navalis*) on the top of a 5-inch-long piece of wood riddled with shipworm tunnels. Note the two small valves at the anterior end of each mollusk that are used for burrowing.

opening when the wood is out of the water. Water retained within the tube allows the animal to survive even after the wood has been out of the water for several weeks. Usually, the animal feeds on plankton and wood, but can survive on wood alone, if necessary. The posterior end of the mollusk remains in contact with the original opening in the wood. That entrance point is never enlarged and the small opening makes it difficult to detect the presence of a shipworm, even when an infestation is extensive. The animals cause severe damage to wooden ship bottoms and pilings; thus, the common name.

Wood pilings are creosoted to protect them against the ravages of shipworms. An untreated piling can be brought to the collapsing point in six months, if heavily infested. Creosoting can prolong the life of a piling to three or four years.

In 1730, shipworms threatened the very existence of Holland by attacking the dikes. They were not identified as mollusks until 1733, when the Dutch zoologist G. Snellius studied the animals causing the damage. In the late 1700s, the hulls of wooden ships were coated with copper sheathing below the water line for protection against shipworms. During the late 1940s, shipworms had disappeared from New York Harbor because pollution had turned the harbor into an unsuitable habitat. That mixed blessing was lost with the 1972 Clean Water Act. By the mid-1980s, the reduced pollution showed positive results. Shipworms were among the various forms of marine life that began to flourish again.

Limnoria lignorum, a very small isopod crustacean that is usually not over ³⁄₁₆-inch long, is often called the gribble or termite of the sea. It also causes considerable damage to wooden structures (Figure 4-11), because it uses mandibles to make a shallow burrow about ¼-inch beneath and parallel to the surface of the wood, around the low-water mark, (Figure 4-12). Burrows can be followed during early infestation

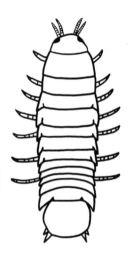

Figure 4-11. The gribble (*Limnoria lignorum*) is a wood-boring isopod crustacean.

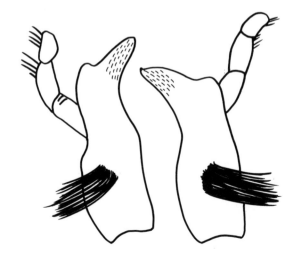

Figure 4-12. Greatly enlarged gribble mandibles.

by a series of small openings that the animal makes to allow a water current to flow through the burrow. Usually a male and female will be present in each burrow, with the female at the blind end, apparently doing all the work. Like all isopod crustaceans, after fertilization the eggs are carried in a pouch under the body. When the 20 or 30 eggs hatch, the young immediately begin boring their own tunnels from the sides of the parents' burrow. The infestation spreads rapidly, and as the outer layers of wood break away, the animals burrow into the lower layers of wood, increasing the depth of damage. As many as 400 animals have been counted in a square-inch piece of wood. The animals swim or crawl to new pieces of wood. Gribbles eat wood fiber, but their main food is a fungus that lives in the wet wood. The gray body is about one-fifth-inch in length with seven thoracic segments and seven pairs of legs with sharp claws, which it uses to anchor itself to the wall of the burrow.

SWIMMING ANIMALS

Wooden piers and docks that extend over the water provide a convenient place to observe free-swimming organisms. Jellyfish are often observed; they are not fish, but cnidarians, members of the class scyphozoa. *Aurelia aurita* is the most frequently encountered jellyfish, often referred to as the moon jellyfish because of its shape and milky white or grayish white color. *Aurelia* has a world-wide distribution. The short tentacles located at the umbrella's margin can cause temporary burning on contact, but the animal is usually considered safe to handle, because the stinging cells are seldom capable of penetrating human skin. Four long, broad, oral arms extend down from the center of the umbrella for use in feeding. As *Aurelia* sinks, plankton becomes trapped in mucus on the undersurface of the umbrella, and ciliary action carries food to the periphery. The frilled oral arms scrape off the food and ciliated grooves pass it to the mouth. *Aurelia*'s four horseshoe-shaped reproductive organs (gonads) can easily be seen through the upper umbrella surface (Figure 4-13). The umbrella may be up to six inches in diameter. A very small polyp is produced; it may live for several years, producing medusae (Figure 4-14). The medusae, so common during the summer, are dead by late autumn, but the next genera-

Figure 4-13. The moon jellyfish (*Aurelia aurita*) has four horseshoe-shaped gonads that can easily be seen through the upper umbrella surface.

tion will be produced by polyps attached to hard substrates below the low tide mark. Figure 4-14 illustrates the alternation of generations which procedes as follows: The medusa (A), the sexual phase of the life cycle, produces sex cells (B) that unite in fertilization. The free-swimming larva (C) attaches to the substrate and develops into a polyp called a scyphistoma (D), the asexual phase in the life cycle. After a period of growth, the tentacles are withdrawn and the body becomes ringed with constrictions. These cut through producing a series of lobed discs, which are young medusae (called ephyrae). The medusae (jellyfish) are released; new tentacles are regenerated and the polyp resumes feeding.

Cyanea capillata, commonly known as the lion's mane or pink jellyfish, is brownish-red to yellow-red and is easily distinguished from *Aurelia* (Color Plate 4a). The umbrella is flat, with eight lobes, and the long tentacles are arranged in clusters of eight. It has four long oral arms. It is often found in swarms during summer months, making the waters uncomfortable for swimmers. The cnidoblasts (stinging cells) are powerful and can easily penetrate human skin, causing painful stings.

The lion's mane is the largest jellyfish in the world. The umbrella grows to about 8 inches in diameter in the study area of this guide, but in Arctic waters the umbrella may attain a diameter of 8 feet, with 800 tentacles. When expanded, the tentacles may be 200 feet in length. It feeds on *Aurelia*, fish and crustaceans.

Dactylometra quinquecirrha is abundant in Narragansett Bay, Delaware Bay, and Chesapeake Bay. The large umbrella, up to 8 inch-

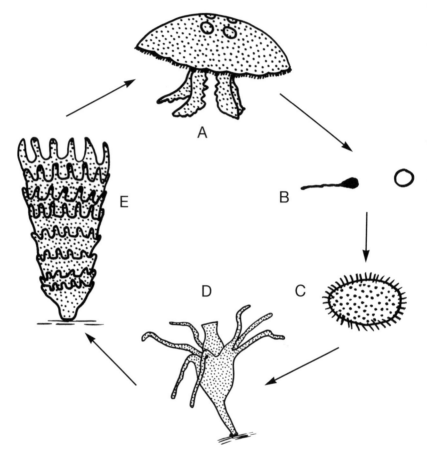

Figure 4-14. *Aurelia aurita* exhibits alternation of generations.

es in diameter, has 48 small flaps of tissue around the margin, giving rise to the common name frilled jellyfish. Raised clusters of cnidoblasts are scattered over the umbrella and they can inflict severe stings. Forty golden yellow tentacles extend down about three feet from the umbrella's margin. Four pink oral arms, as long as the tentacles, surround the mouth on the undersurface.

The purple jellyfish (*Pelagia noctiluca*) is so named because of the umbrella's color. The umbrella is usually about two inches in diameter with red dots scattered over the upper surface. The animal has four long, oral arms extending from the center of the undersurface and about eight tentacles trailing from the umbrella's margin.

The victim of a jellyfish sting should remain quiet and in a comfortable position to decrease muscular activity that increases the circulation of venom through the body. It has been reported that the burning sensation of jellyfish stings may be neutralized by applying Burrow's solution (available in most drug stores) to the affected area. In the waters south of Chesapeake Bay, a good rule of thumb for neutralizing jellyfish nematocysts (the stinging component of a cnidoblast) is to apply acetic acid (vinegar) in liberal amounts to the affected area. North of Chesapeake Bay, applying a mixture of 50% water and 50% baking soda to the injury is recommended. These measures do not always stop the pain and swelling, and a physician who can make a continuing diagnosis of the patient's condition should be contacted.

Often, identification of the species of jellyfish can make treatment easier. If you do not know the species of the "gelatinous offender," a sketch of the animal, or at least documentation of the location, time, and water conditions where the incident occurred, may be helpful.

Ctenophores (comb jellies) resemble jellyfish; they are often seen just beneath the surface, especially when the water is calm. Also, they are often washed ashore as gelatinous blobs. Ctenophores in the study area do not possess cnidoblasts and cannot sting. They feed by swimming with the mouth forward so plankton can be swept in, or by trailing a pair of sticky tentacles. Some species of fish feed on ctenophores.

Mnemiopsis leidyi, commonly called the sea walnut, is the most frequently encountered ctenophore in the study area covered by this guide. It has a moderately flattened, transparent body that forms two large lobes. The two lobes and the presence of comb rows distinguish the sea walnut from cnidarian medusae. The comb row has tiny hairlike projections like the teeth of a comb. The rapidly beating combs that are used for locomotion refract light and produce an iridescence. The name ctenophore means "comb-bearer." Short tentacles border the slitlike mouth of the filter feeder, which grows to about four inches in length. It is found south of Cape Cod, often in dense groups called swarms, during summer months. When disturbed or irritated, the sea walnut can produce bioluminescence, an enzyme-catalyzed chemical reaction that results in flashes of greenish-white light. When the animal, carried by a wave, strikes a piling or is disturbed by a passing boat, it produces the light.

Another organism that produces light when disturbed by a boat's wake or waves breaking along the shore is the unicellular *Noctiluca*. Unicellular animals and plants belong to the kingdom protista and are not covered in this guide, which covers macroscopic not microscopic organisms. The genus name *Noctiluca* means "light of the night"; on dark, moonless summer nights its pin-point flashes of light are often referred to as "star phosphorescence." By splashing your hand in the water some dark night, you may see star phosphorescence. Apparently, bioluminesence serves no real function in *Noctiluca* or comb jellies.

Luminescent, microscopic bacteria (kingdom monera, also not covered in this guide) are present in the ocean and can cause decaying fish to glow in the dark. Many species of invertebrates produce bioluminescence. Among vertebrates, luminescence is found only in certain fish.

The ctenophore sea gooseberry (*Pleurobranchia pileus*) has two tentacles covered with adhesive cells to trap plankton and fish eggs (Figure 4-15, and Color Plate 5a). The tentacles then wipe the food into the mouth. It grows to about one-inch long, and is usually found in deeper water, but can occur in large numbers inshore in the winter and spring.

Figure 4-15. The sea gooseberry (*Pleurobranchia pileus*) ctenophore has two long tentacles.

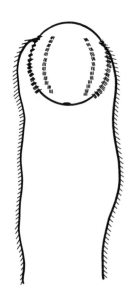

Late in the day and at night, around well-lighted docks, squid, cephalopod mollusks, may be seen darting about in search of prey, especially in the spring, when females deposit egg masses, called mops, in shallow water. The slender, white masses, about two inches long, are laid in communal clusters of about 40 fingerlike structures, each containing about 100 eggs (Color Plate 8d). The egg masses must have a bad flavor, as they are not preyed upon by other marine organisms.

In squid, the internal chitinous shell, called a pen, is reduced and embedded in the mantle. The pen is a supporting structure for the fast, streamlined swimmer. Eight short arms, with two rows of suckers and two long tentacles with terminal clubs, are characteristic. The body is elongate or torpedo-shaped, with a pair of triangular fins at the pointed end of the mantle. Whales, porpoises, seals, and the larger and faster fish prey on squid. Squid produce a black ink that is ejected when the animal is threatened by a predator. The "smoke screen" confuses the sight and smell of the attacking animal while the squid escapes. Ink recovered from squid ink sacs is sold as India Ink, although synthetic substitutes have almost completely replaced that natural ink. A flashlight held over the water at night may attract squid, crabs, and other animals.

Loligo peali is the common squid in the study area of this guide. Body length, excluding arms and tentacles, averages four to five inches. The paired fins are at least half the length of the mantle.

Loligo brevis is a smaller species, body length less than three inches, found from Delaware Bay southward. The triangular fins are shorter and do not extend halfway down the mantle (Color Plate 8c).

If you catch a live squid, its brown and purple pigment cells (chromatophores) will put on an impressive show. The pigment cells of the irritated animal pulsate, enlarging and contracting, quickly changing its body color. The expansion or contraction of the pigment cells is controlled by fine strands of muscle attached to the outer wall. Normally the pigment cells are camouflage, allowing the squid to change its body color to blend in with different surroundings.

SHIP-FOULING COMMUNITIES

These communities are marine organisms attached to the hulls of ships. Attachment of marine plants and animals is least where there is much fresh water or much pollution. Settlement only occurs when a ship is stationary in port, but the organisms continue to grow while the ship is moving. In temperate seas, such as the study area of this guide, it has been estimated that the frictional resistance due to the fouling organisms increases by one-quarter of one percent per day. C. M. Yonge in *The Sea Shore* states, "six months after it leaves dock, the maximum speed of a battleship of 35,000 tons is reduced from this cause by one and one-half knots while an additional 40% of fuel is needed to maintain a speed of twenty knots. On the average probably 20% of the fuel used by a ship is needed to overcome the added resistance due to the growth of this encrusting life."

Ship-fouling communities continuously introduce non-native species. The European periwinkle (*Littorina littorea*) was either introduced intentionally by the colonists or in rock ballast. In the latter part of this century some organisms arrived in our waters through the ballast water of ships. Plankton and the larval form of some organisms can survive in the rather inhospitable environment of ballast water. In 1993, a research group at the Smithsonian Environmental Research Center in Edgewater, Maryland, found that a diverse biota of plankton and fish are present in ballast water. The researchers estimate that 1.7 million gallons of plankton ballast water are released in United States waters from international vessels every 60 minutes. The European bryozoan *Membranipora membranacea* and the European sea squirt *Botryllus schlosseri* were believed to have been introduced in this manner. The invasion of new species continues today. The invaders compete with, and often displace, native species and affect predator-prey relationships.

A SIMPLE EXPERIMENT

You can study the progression of the attachment of marine organisms to wood with a flat piece of wood. Weight it at one end and cut a hole in the other end. Tie a strong piece of string through the hole and tie the other end to a dock in the quiet waters of a bay. Then, drop the wood into the water as early as possible in the spring. Pull the piece of wood from the water once a week and inspect its surface with a hand lens. After several months, a community of algae, barnacles, hydroids, bryozoans, and other animals should be evident.

5

Bays and Estuaries

But I have sinuous shells of pearly hue . . .
Shake one, and it awakens; then apply
Its polished lips to your attentive ear,
And it remembers its august abodes,
And murmurs as the ocean murmurs there.

Walter Savage Landor

Wave-driven currents along the shore usually remove mud from beaches and carry it along in suspension. The calmer waters of quiet bays and estuaries allow the mud particles to settle, producing bottoms and beaches that vary in composition; some areas having a predominance of sand while other areas are sand and mud or mainly mud, often called mudflats (Color Plate 1c).

The sand and mud or mud sediment provides a more stable habitat for marine organisms than a sand beach. Beach sand is fairly coarse, and water flows through it more readily. Fine particles of mud pack together, and are less penetrable. The finer the mud particles, the denser they pack, permitting less water and oxygen to pass through to the deeper layers.

Digging into a mud bottom often produces the rotten egg smell of hydrogen sulfide. At the surface, the sediment is usually gray because oxygen is present, and aerobic bacteria break down most of the organic matter. A black layer usually lies just under the surface. There is little or no oxygen penetration below the black layer, and few organisms can survive there. However, anaerobic bacteria function without oxygen; they continue to break down organic matter containing sulfate ions, reducing it to hydrogen sulfide in the process.

BURROWING ANIMALS

The bottom, when exposed at low tide, may seem barren, but many marine animals are burrowers. There are relatively few predators in the sediment, but surface predators can often reach into it. An octopus can run a tentacle down a burrow, and a fish such as the sting ray can excavate the sediment by movement of its wing-like fins. The sand star (*Astropecten* species) has suckerless tube feet, thus it cannot climb rocks like sea stars. But the tube feet are pointed, and adapted for digging. A sand star can burrow rapidly through the sand to prey upon crustaceans, mollusks, annelids, and other echinoderms. It is relatively large, eight to ten inches in diameter. The color varies, but is usually orange. Occasionally, the upper surface is purple with broad, orange-red marginal plates and purple spines (Figure 5-1). The sand star does not have pedicellariae. It is found from Long Island southward, rarely north of Long Island.

At low tide, when the mudflats are exposed, birds with long beaks can reach the burrowers.

Some mollusks, such as clams, burrow into the bottom and use their siphons to obtain food and oxygen from the clean water above the sediments (Color Plate 8b). Bivalves, with the lateral compression of the shell, are perfectly adapted for burrowing in soft sediments. The original name for the class to which clams belong is *Pelecypoda*, which means "hatchet foot" in Greek. The pointed tip of a clam's muscular foot is its burrowing tool. It is inserted into the bottom, then the tip is inflated with blood, forming a hatchet shape that anchors the animal in the sand or mud. As the strong muscles within the foot contract, the mollusk is pulled into the bottom in a process that is repeated until the clam has buried itself.

The familiar clam *Mercenaria mercenaria* is of great economic importance, annually accounting for millions of dollars of commerce. The scientific name means "money-conscious." Among the many common names used today are quahog and hardshell clam. The name quahog

Figure 5-1. The sand star (*Astropecten* sp.) is distinguished by wide upper and lower marginal plates. The specimen on the left, showing the aboral surface, is 2.7 inches across.

(pronounced co-hog) was given to these mollusks by the Native Americans (Indians); the white settlers called them clams, the Anglo-Saxon word for clamp. Small ones, called little necks, are up to one and a half inches long; they are from three to four years of age and are highly favored as clams on the half shell. Mid-sized ones, called cherrystones, are two inches long, probably five years old, and are commonly used for clambakes. Large ones, from six to fifteen years old, are called chowders. Quahogs can grow to over four inches, with thick, oval-shaped shells. Many conspicuous growth rings are present, especially near the margin on their unattractive, grayish exterior surface. The interior shell color is white, with purple margins and markings (Figures 5-2 and 5-3).

Parts of quahog shells were used as wampum (currency) by Native Americans. The shells were sectioned, pierced, and rubbed between stones to form cylindrical beads about one inch long and one-eighth inch in diameter. After polishing, the beads were strung together as strands or belts. Purple beads made from the inside of the shell were twice as valuable as those from the white portion. Occasionally, whelks and periwinkles were used to produce wampum. Native Americans on the Pacific coast used tusk shells as their currency.

Figure 5-2. The quahog (*Mercenaria mercenaria*) is of great economic importance. The specimen is three inches long.

Figure 5-3. Valves of a quahog (*M. mercenaria*), with a peeling, thin, black periostracum.

Quahogs, like many bivalves, can secrete calcareous pearls, but only those species that contain iridescent nacre, the innermost layer of shell known as mother-of-pearl, can produce pearls of commercial value. *Mercenaria* shells do not contain nacre; their white or purple pearls lack luster and have no real value, although parts of the shell are often used in the production of jewelry.

Brackish water, a combination of fresh and salt water, is preferred by *Mercenaria*, which is why they are found in estuarine and bay waters rather than the ocean. They protect themselves in winter months by burrowing deeper into the sediment. Long-handled rakes with curved tines, or tongs with long wooden handles are used to gather clams. They are served cooked or raw, but care must be exercised to avoid eating those from polluted waters, particularly when they are raw. The filter-feeding animals may contain pathogenic organisms from their polluted habitat.

Another clam collected for its food value is *Mya arenaria*, commonly called steamer, soft-shell clam, or long neck clam. The shells are grayish white, thin, elliptical, and about three inches long. The valves gape at the ends, and do not close tightly. A large spoon-shaped tooth is located in the hinge region of one valve (Figure 5-4). The two

Figure 5-4. The hinge ligament of the soft-shell clam (*Mya arenaria* is partially internal and is carried on a conspicuous, spoon-shaped shelf that projects from one valve and underneath the umbo of the opposite valve. The specimen is 2.75 inches long.

siphons are fused to form a long structure that cannot be completely retracted between the valves, and gives rise to the common name long neck clam. The pallial sinus (Figure 1-18) is large and U-shaped. This familiar clam, although popular as a food, is not as important commercially as *Mercenaria*. The clam burrows into soft bottoms, especially in the intertidal zone, leaving a finger-size hole flush with the surface of the sand. Placing your foot by the opening and applying pressure will usually produce a squirt of water as the clams withdraw their siphons. In early life *Mya* moves about on the bottom and even makes temporary attachments with byssal threads. But with increasing size it begins to burrow beneath the surface.

Large, wedge-shaped holes flush with the sand may indicate the presence of razor clams (*Tagelus plebius* and *Ensis directus*), so named because the shell is shaped like an old-fashioned straight razor. When submerged, razor clams are just beneath the surface of the sand with only a small part of the shell exposed, but as the tide recedes, they burrow deeper. They are the best-flavored of all the clams. Shovels are used to collect them, but like a magician with a disappearing act, they often evade capture by using the muscular foot to rapidly burrow deeper into the bottom. The narrow, even-width shell facilitates burrowing and they are the fastest diggers of the tidal flats.

As described earlier, the tip of the muscular foot is the burrowing tool. When the muscles within the foot contract to pull the shell deeper, the shell valves are drawn closely together by adductor muscles. When the foot is ready to be extended again, the valves separate to anchor the clam in its temporary position. If they are not captured in the first shovelfull of sand, you might as well look for another wedge-shaped hole in the sand.

Tagelus is commonly called blunt razor clam, stout razor clam, and stout *Tagelus*. It has a slightly inflated, stout, oblong shell. Shell length is about three to four inches. Concentric lines wrinkle the smooth surface. Color is white or yellowish. *Ensis* is often called the common razor clam or jackknife clam. It can easily be distinguished from the blunt razor clam by its long (over seven inches) slightly curved, thin shells (Figure 5-5). The shell is white with a greenish-brown or yellowish-brown periostracum.

Figure 5-5. A 2.5-inch-long stout razor clam (*Tagelus plebius*) is on the left, note the extended siphons. The valves of the jacknife razor clam (*Enis directus*) in the center are open to show the mantle and fingerlike foot. The jacknife razor clam on the right is 4.25 inches long and has the tip of its foot extended between the valves.

The growth rings of *Astarte castanea* are not as prominent as other astartes. Thus, it is commonly called the smooth astarte. The bivalve has a small, approximately one-inch-thick, shell with a curved beak. Each valve's inner margin is crenulated. Exterior color is reddish brown, interior color is white. Because of the chestnut external color, it is often called the chestnut astarte. The bivalve's body is bright red, like that of blood arks.

Spisula solidissima, commonly known as the surf clam or hen clam is the largest bivalve in the study area, up to seven inches in length. As the common name implies, the bivalve is found in sand of the surf zone and in deeper waters. The valves are thick, somewhat triangular-shaped, and have a large, centrally-located umbo. There is a large spoon-shaped cavity for the ligament in the hinge area (Figure 2-44). The valves are smooth, with fine growth rings and thin, brown periostracum. Color is yellowish white. The commercially important clam is very popular for clambakes, accounting for 70% of the clams harvested in the U.S. Only the adductor muscles are eaten.

An anemone, *Haloclava producta*, burrows into the bottom, with only its tentacles exposed, and is known as the burrowing anemone. The column is approximately three inches long, and the upper part is covered with rows of small round protuberances. There are 20 tentacles with enlarged tips.

The striped sea anemone (*Haliplanella luciae*) is occasionally encountered in the bay environment, but does not burrow into the bottom. It has a green to brown column that often has vertical stripes ranging in color from white to red. There may be as many as 50 tentacles in several whorls. The column of this small anemone seldom exceeds one inch in length when expanded. It is often found on the mud of tidal marshes, or attached to rocks or pilings. This species, an immigrant from Europe, appeared first in Connecticut about 1892 and has since spread throughout the study area.

Although most beachcombers are not excited by worms, many very colorful annelids are found in this habitat. Those that burrow continuously are usually predators searching for prey. Others that make permanent burrows are usually detritus or filter feeders. Deposit feeders are often equipped with tentacles covered with fine hairs called cilia; they accept only certain kinds of detritus and reject other particles. An example is the plumed worm (*Amphitrite ornata*). It builds a mucus-lined tube of sand and mud. The head bears many long extensible, flesh-colored tentacles that gather food. Ciliated grooves pass the food to the mouth for ingestion. The tentacles are also used to gather sand for the tube. Three pairs of bright red, highly branched gills are located just behind the tentacles. This worm can grow to 15 inches in length; its body, except the gills, is flesh-colored. It is frequently found from New Jersey to Vineyard Sound by looking for tiny volcano-shaped mounds on the tidal flats.

Other deposit feeders are not as selective. They simply ingest large quantities of sand or mud and excrete what is useless. The lug worm (*Arenicola cristata*) is an example. The head is moderately developed, without tentacles, but possessing eyes. Parapodia are small and not well developed. Thirteen pairs of brownish-red tufted gills are located in the middle of the greenish blue-black body. These worms are six to eight inches in length and as thick as a pencil.

Arenicola constructs an L- or U-shaped burrow in the lower intertidal zone, and strengthens the wall with mucus. The burrow is identifiable by the characteristic mounds of coiled castings (a mixture of feces and sand grains) at the back end (Figures 5-6 and 5-7). Muscular contractions of the worm's body pulls water through the burrow. In late spring it produces large balloon-like clear gelatinous egg sacs, containing thousands of tiny reddish-brown eggs. It is even found in brackish water. The polychaete is often used as bait.

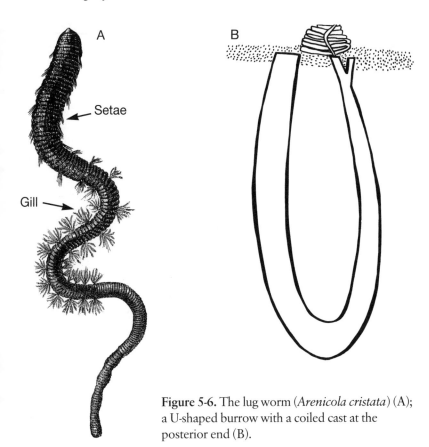

Figure 5-6. The lug worm (*Arenicola cristata*) (A); a U-shaped burrow with a coiled cast at the posterior end (B).

Figure 5-7. A lug worm's coiled cast on a mudflat. The ribbed mussel (*Modiolus demissus*) is about 2 inches long.

Some worms, while burrowing, simply push the sand or mud aside with their pointed heads. Lug worms, however, burrow by means of a proboscis, a muscular protuberance from the head, that can be extended and inflated in much the same way as a clam uses its muscular foot.

The parchment worm (*Chaetopterus variopedatus*) lives in a U-shaped parchment tube, secreted by its own body, with the ends projecting like chimneys above the sand. It has an interesting method of gathering food. The parapodia on three of its middle segments are shaped liked fans. Each pair forms a sort of piston that fits against the cylindrical wall of the tube. Movement of the fans, about 60 times a minute, produces a front-to-back current of water through the tube, bringing in oxygen and food. The long, wing-like parapodia in front of the fans secrete a mucus film, resembling a net stretched between two poles. Plankton is filtered out as water passes through the mucus net. The end of the continuously secreted net is attached to a cup-like structure that rolls up the film. When the ball of film reaches a certain size, the fans stop pumping and the ball is passed along a ciliated groove to the mouth. Large objects brought in with the current that might damage the net are detected by cilia, and shunted aside. The wing-like parapodia are then raised, to avoid damage to the net. The Latin specific name refers to the various shaped parapodia in different regions of the body. The worm's body can be about ten inches in length and is luminescent, which seems strange because it lives out of sight within a tube. (Figure 5-8).

Figure 5-8. *Chaetopterus variopedatus* in its U-shaped parchment tube.

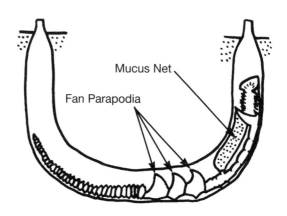

Mucus Net

Fan Parapodia

The sand worm or clam worm (*Nereis* species) often lives in a burrow from which its tentacles protrude. Small animals passing near the opening are snatched by the worm's jaws and devoured. There are several species of *Nereis*. *Nereis virens* is one of the largest and most common annelids found in the study area. It is distinguished by a well developed head, with a pair of short tentacles on the upper side of the anterior end between two large stubby, jointed palps, and two pairs of longer tentacles on each side of the head. The tentacles are sensory receptors for touch and the palps possess sensory receptors for taste and smell. The worm has four small dorsal eyes. It is an iridescent mixture of blue and green, the tentacles have a touch of red, and its limb-like parapodia are usually orange. A short muscular proboscis with a pair of chitinous jaws can be thrust from the mouth to seize worms and other invertebrates (Figure 5-9). The proboscis is retracted and the powerful jaws tear the prey to pieces. The clam worm also feeds on plant detritus. Large specimens are approximately 18 inches long, with about 200 segments. Each segment has a pair of lobed parapodia.

During the day, *N. virens* can be found in a mucous-lined burrow in the sand, hidden among algae, or under rocks in the intertidal zone or

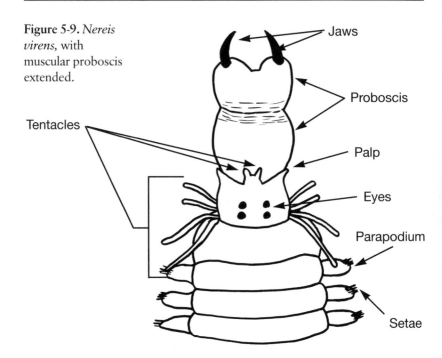

Figure 5-9. *Nereis virens,* with muscular proboscis extended.

sublittoral. The limblike parapodia, with chitinous bristles (setae) at the end, help to anchor the worm in its burrow (Figures 5-10 and 5-11). At night, the voracious predator swims about like an eel in search of food (Color Plates 9a and 9b). The parapodia, used as paddles, aid in locomotion. After a night of foraging for food, the clam worm returns to the same burrow, using special sensors in the tentacles to find it. It feeds on young clams and mussels, other worms, and crustaceans, and in turn is preyed upon by fish and crabs. The annelid is frequently sold for bait, but its large jaws can give a painful bite to an unwary fisherman or beachcomber. During the breeding season, clam worms congregate near the surface of the water.

Glycera americana is known as a beak thrower because its large, clublike pharynx can be quickly everted to capture prey with four beaklike, curved, black hooks on the end. Each hook has a duct connected to a poison gland running through it (Figure 5-12). The translucent, light purple body has a pink line running dorsally through its entire length and small pink parapodia. The worm, that grows to a foot

Plate 9a. A juvenile clam worm (*Nereis virens*) swimming in the water column. (Photo by Norman Despres.)

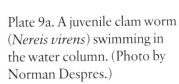

Plate 9b. The head of an adult clam worm. (Photo by Dave and Sue Millhouser.)

Plate 9c. Living and dead (open valves) acorn barnacles (*Balanus* sp.) on a rock. The blue mussels (*Mytilus edulis*) on the left are only .25-inch in length.

Plate 9d. Acorn barnacles on the stipe of a kelp, with their feeding appendages extended. (Photo by Norman Despres.)

Plate 9

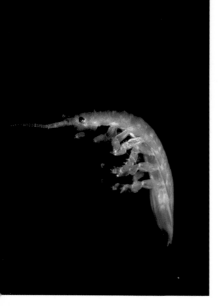

Plate 10b. A skeleton shrimp (*Caprella* sp.), an amphipod with a bizarre skinny body, stands erect on an alga. (Photo by Dave and Sue Millhouser.)

Plate 10a. An amphipod crustacean. (Photo by Dave and Sue Millhouser.)

Plate 10c. A shrimp (*Crangon* sp.), a decapod crustacean, blends into the background. (Photo by Dave and Sue Millhouser.)

Plate 11a. A cast of a horseshoe crab (*Limulus polyphemus*) in the wrack line. The old exoskeleton is split in the front, between the upper and lower carapace. The animal emerged through this opening, unlike true crabs that molt from the rear. (Photo by Theano Nikitas.)

Plate 11b. The ghost crab (*Ocypode quadrata*). (Photo by George Harrison, courtesy of the U.S. Fish and Wildlife Service.)

Plate 11c. A ghost crab about to escape into its burrow. (Photo by Theano Nikitas.)

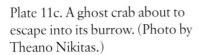

Plate 11d. A spider crab (*Libinia emarginata*) has camouflaged itself with algae. Note the eye above the tip of the crab's right pincer claw. (Photo by Jim Matulis.)

Plate 11

Plate 12a. A dead blue crab (*Callinectes sapidus*). (Photo by Theano Nikitas.)

Plate 12b. A broad-clawed hermit crab (*Pagurus pollicaris*) with the hydroid snail fur (*Hydractinia echinata*) on its shell. Note the small long-clawed hermit crab (*Pagurus longicarpus*) on the left. (Photo by Dave and Sue Millhouser.)

Plate 12c. A broad-clawed hermit crab with a tortoise shell limpet (*Acmea testudinalis*) just above the base of the crab's right tentacle. (Photo by Dave and Sue Millhouser.)

Plate 12d. A male broad-clawed hermit crab holds onto the gastropod shell of a female. (Photo by Dave and Sue Millhouser.)

Plate 13a. A green crab (*Carcinus maenas*).

Plate 13b. The eastern sea star (*Asterias forbesi*). Note the orange madreporite. (Photo by Dave and Sue Millhouser.)

Plate 13c. A moon snail (*Lunatia heros*) moves over Atlantic sand dollars (*Echinarachnius parma*). (Photo by Michael deCamp.)

Plate 13

Plate 14a. Green sea urchins (*Strongylocentrotus droebachiensis*) cover the rocks in some areas. (Photo by Norman Despres.)

Plate 14b. Green sea urchins on a brown alga (*Laminaria* sp.). Note the extended tube feet of the one on top. (Photo by Norman Despres.)

Plate 14c. A close-up of a purple sea urchin (*Arbacia punctulata*). (Photo by Chip Cooper.)

Plate 15a. The colonial star tunicate (*Botryllus schlosseri*). (Photo by Jim Matulis.)

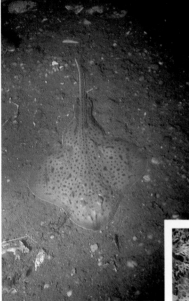

Plate 15b. The skate *Raja eglanteria* has a ridge of thornlike spines and dark brown spots on its back. (Photo by Michael deCamp.)

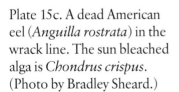

Plate 15c. A dead American eel (*Anguilla rostrata*) in the wrack line. The sun bleached alga is *Chondrus crispus*. (Photo by Bradley Sheard.)

Plate 15d. Black skimmers (*Rhynchops niger*). Note the longer lower mandible. (Photo by Jack Darrell, courtesy of the U.S. Fish and Wildlife Service.)

Plate 15

Plate 16a. Common egret (*Casmerodius albus*). (Photo by Luther C. Goldman, courtesy of the U.S. Fish and Wildlife Service.)

Plate 16b. Great blue heron (*Ardea herodias*). (Photo by John Cossick, courtesy of the U.S. Fish and Wildlife Service.)

Plate 16c. Brown pelican (*Pelacanus occidentalis*). (Photo by Mike Haramis, courtesy of the U.S. Fish and Wildlife Service.)

Plate 16d. American oystercatcher (*Haematopus palliatus*). (Photo by Luther C. Goldman, courtesy of the U.S. Fish and Wildlife Service.)

Plate 16

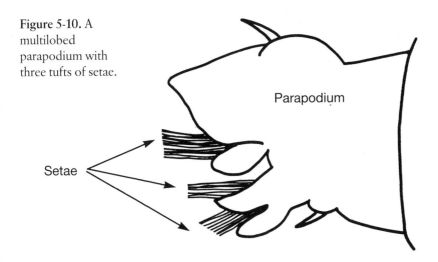

Figure 5-10. A multilobed parapodium with three tufts of setae.

Parapodium

Setae

Figure 5-11. Part of a parapodium with a tuft of setae, magnified 40 times.

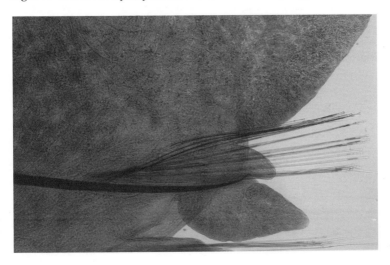

in length, can be found in burrows in sand or mud flats. When burrowing, the animal rotates its body, literally screwing itself into the bottom. It lies in wait within the burrow and can detect prey by changes in water pressure. If a worm is removed from its burrow, it will repeatedly extend and retract the pharynx.

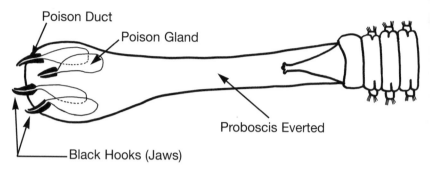

Figure 5-12. The pointed head identifies the beak thrower *Glycera americana.*

G. americana is also known as the bloodworm because red blood can be seen pulsing through the translucent body. It is sold for bait and is found throughout the study area.

The body segments of *Clymenella torquata* are longer than they are wide. The characteristic long body segments of this worm have given it the common name bamboo worm (Figure 5-13). Its poorly developed head is truncated, and the terminal segment is funnel-shaped with scalloped edges. Parapodia are rudimentary. The tube-dwelling worm constructs a long, straw-size, round, straight-sided tube of sand grains and mucus that extends above the bottom. It is approximately six inches in length and light red in color, with bright red bands. It feeds on detritus and is common in the lower intertidal areas to deep water, from New Jersey northward.

Polydora ciliata is a tube-dwelling annelid commonly known as the tentacled worm because it has a pair of long tentacles on its head. The fifth body segment is larger than the others. It is a filter feeder and lives in a small upright tube, often in large colonies on the sandy or muddy bottom. The tentacled worm is about one inch in length. Body color is pale green, gray, or orange with white tentacles. It is usually found at the low-tide mark.

Diopatra cuprea is commonly called the decorator worm because it builds a long parchment tube camouflaged with sand, bits of shell, and seaweed (Figure 5-14). The tube has a single chimney with the top at a 90° angle. It has a well developed head with one long pair of tentacles, one short pair of tentacles, one pair of slender palps, and a pair of ocelli (light sensitive areas). Bright red, feather-like gills are paired on the

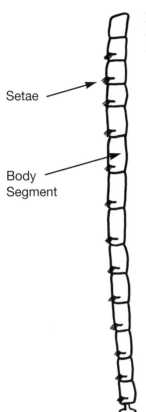

Figure 5-13. The longer than wide body segments of *Clymenella torquata* are responsible for the common name bamboo worm.

Setae

Body
Segment

Figure 5-14. A 7-inch-long tube of the decorator worm (*Diopatra cuprea*). Note the shells attached to the parchment tube.

anterior segments giving another common name, plumed worm, to the beautiful annelid. It is a predator that partially extends its body, over 12 inches in length, out of the tube to seize tiny invertebrates. It is common from Long Island Sound southward.

Pectinaria gouldii constructs a cone or trumpet-shaped sand-grain tube, open at both ends, with the large end down. Thus, it is commonly known as the cone worm, ice cream cone worm, trumpet worm, and mason worm. The delicate cone is composed of a single layer of closely fitting sand grains glued together and occasionally washes ashore. The largest grains of sand are positioned at the wide end of the tube and smaller ones at the pointed end. Only the pointed end extends above the sand (Figure 5-15). After feeding on sediment at one spot for several hours, it moves to another area and burrows head first into the bottom. The cone worm's head, at the end of the trumpet-shaped body, bears a tuft of short tentacles and longer, golden-yellow bristles. Well-developed, paired parapodia are present on most body segments. The pink-colored worm reaches a length of about two inches. The cone worm is often found by looking for a black smudge (feces), about one-half-inch in diameter, on the sand or mud.

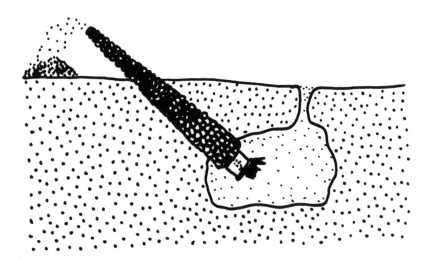

Figure 5-15. The ice cream cone worm (*Pectinaria gouldii*) constructs a tube of grains of sand that are carefully selected for size and fitted together with precision. The specimen is 2 inches long.

The uniform appearance of the sand or mudflat surface may be interrupted by piles of castings made by the worms mentioned previously and other species. Many are deposit-feeders that eat large amounts of sediment, remove the organic material during the digestive process, and then deposit the residue in the form of castings at the back end of their tube or burrow. The many piles of castings and openings of burrows and tubes on the surface of mudflats attest to the abundance of animal life beneath (Figure 5-16). During the summer, mudflats are spotted with gelatinous masses of annelid eggs. The egg cases are of varied shape and often are mistaken for jellyfish, but close inspection reveals hundreds of small red or white dots within the gelatinous mass.

Some species of shrimp burrow into the bottom. The mantis shrimp (*Squilla empusa*) is shrimplike in appearance, but the body is broad and flattened dorsoventrally. The common name is derived from the long front legs with scythelike claws, similar to those of a praying mantis. The long claws are called a "jacknife claw," because the outer blade fits

Figure 5-16. A stainless steel and copper strainer, and a folding shovel can be used to collect tube worms and other buried marine organisms.

into a groove in the inner part like the folding blade of a jacknife (Figure 5-17). The large, to ten inches, crustacean digs a burrow in muddy sand near and below the low-tide line, and waits at the entrance to pounce on small fish and invertebrates. The slashing movement of the claws "can cut a shrimp in two as though it had been guillotined." The claws can inflict a painful cut to a beachcomber if handled. Shell coloration is greenish, edged with bright yellow.

The large claws on the small, to two and a half inches, mud shrimp (*Callianassa atlantica*) give it a lobsterlike appearance (Figure 5-18). The shrimp burrows in sandy mud in the lower intertidal zone and sublittoral, and is found throughout the study area. Burrows extend down about one and one-half feet and are U-shaped with two entrances.

Some echinoderms (sand dollars, sea cucumbers, and sea stars) may burrow into the bottom, but usually just beneath the surface. The short-spined brittle star (*Ophioderma brevispina*) is usually found in bays and

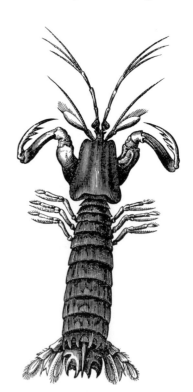

Figure 5-17. The body of the mantis shrimp (*Squilla empusa*) is broad and flattened dorsoventrally. The crustacean has scythe like claws.

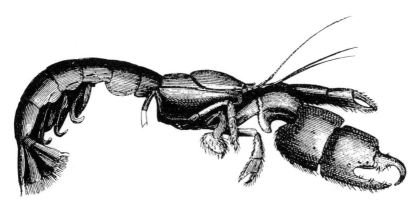

Figure 5-18. The large claws of the mud shrimp (*Callianassa atlantica*) give it a lobsterlike appearance.

estuaries, and is common in the northern region of this study area. It is sometimes found under rocks and is easily recognized by the seven or eight short spines on the side of each arm segment (Figure 5-19). As the common name indicates, the animal is fragile and the arms are easily broken. The color ranges from olive-green to black. The central disk can be three-quarters of an inch in diameter, the arms about four times longer.

Figure 5-19. The short-spined brittle star (*Ophioderma brevispina*). The central disk is .5-inch in diameter.

An echinoderm that is often impaled on the tines of clam rakes as bivalves are being collected is the sea cucumber. The animal, when disturbed, such as being pierced by the tines, retracts its tube feet and feeding tentacles. Sea cucumbers do not have spines like the more typical echinoderms, such as sea stars and sea urchins. With the tube feet retracted it is difficult to recognize the animal as an echinoderm.

The most frequently encountered sea cucumber in the study area is *Sclerodactyla briareus*. Its elongated body is covered with tube feet giving rise to the common name hairy cucumber (Figure 5-20). It is usually brownish-black in color.

Another echinoderm, the glass sea cucumber (*Leptosynapta* species), usually burrows into the mud. When mature, it has a white or pink tinted, transparent body about six inches in length. The fragile animal does not have tube feet, but does have the sea cucumber's characteristic feeding tentacles.

Figure 5-20. The sea cucumber *Sclerodactyla briareus*; extended tube feet give rise to the common name hairy cucumber. Note the feeding tentacles on the right. The specimen is 2 inches long.

NON-BURROWING ANIMALS

At low tide the crisscrossing groove-like trails of snails can be seen on the exposed mudflats. One is the mud snail (*Ilyanassa* [*Nassarius*] *obsoleta*) (Figures 5-21 and 5-22). It is the most abundant gastropod in the study area, often congregating in packed masses that can temporarily cover acres. The shell is thick, with an elevated spire, which is usually covered with algae and worn down at the end in older snails. Average shell length is about 3/4-inch and has approximately six whorls, marked with unequal revolving lines of flattened beads crossed with small growth lines. Shell color varies from reddish brown to purple-black. The snail crawls about, actively feeding on algae and organic material in the mud. The snail is also a scavenger; it feeds on dead animals, which it is able to detect from great distances with its sense of smell. It is not unusual to find a dead animal covered with mud snails.

Mud snails are necessary to complete the life cycle of a flatworm (phylum platyhelminthes, not covered in this guide), the blood liver fluke (*Cercaria variglandis*), the causative organism of "swimmer's itch." Blood liver flukes live in shore birds and their eggs pass out in

Figure 5-21. The mud snail (*Illyanassa obsoleta*). The specimen showing the spire is .8-inch-long. Note the eroded spire and, inside the aperture, the longitudinal folds on the outer lip.

Figure 5-22. A mud snail (*Ilyanassa obsoleta*) leaves a conspicuous trail on a mudflat. The valves of a blue mussel (*Mytilus edulis*) are below the snail.

the bird's feces and hatch into microscopic ciliated forms that are eaten by mud snails. The ciliated forms grow and develop into larvae that are released by the snail. Normally, in the spring, the larvae would penetrate the skin of a shore bird to repeat the cycle. However, if a human is present the larvae can penetrate the skin, where they die. Human skin is too thick to allow the larvae to reach blood vessels, but the dead larvae cause an irritation that itches like poison ivy.

Another gastropod, the basket snail (*Nassarius trivittatus*), is frequently seen. It can be easily distinguished from the mud snail by its suture channels and the prominent beads on its shell (Figure 5-23). The revolving lines of beads are crossed by growth lines, giving a basket-like appearance to the shell. Exterior color varies from white to yellowish white, with purple-brown mottling or stripes.

Carnivorous snails, such as the dog whelk (*Thais* species), that are numerous on rocky shores are also found in this habitat. The dog whelk will bore a hole through the shell of another mollusk, even another dog whelk, to get at the soft body tissue within. The snail is described in Chapter 3, the "Rock Beaches and Jetties."

Figure 5-23. The shell of the basket snail (*Nassarius trivittatus*) has a basket-like appearance. The specimen on the left is .85-inch-long.

The empty shells of gastropods that are usually found below the low-water mark are frequently washed up onto the beach. A few examples are:

Crepidula fornicata is commonly known as the slipper shell, quarterdeck, or boat shell. The shell differs from that of other snails in that it has a shelf inside. The shell's cavity has a plate that covers about half the concave interior, giving it the appearance of a slipper or the quarter deck of a vessel (Figure 5-24). The slipper shell is oval, arched, and grows to about 1½ inches in length. Shell color is white with pinkish or purplish spots and stripes. The white interior shelf covers about one half of the aperture. *Crepidula* species are suspension feeders, filtering out plankton with tremendously lengthened mucus-covered gill filaments. The trapped plankton is moved, by ciliary action, toward the mouth where the radula seizes it and pulls it into the digestive tract. *Crepidula* tend to live stacked on one another, with older, larger shells on the bottom. Each animal in a stack is believed to be one year younger than the one to which it is attached. Reproduction is interesting because young specimens are always males, but this initial phase is

Figure 5-24. The cavity of a slipper shell (*Crepidula fornicata*) has a plate that covers about half the concave interior. The specimen on the left is 1.5 inches long. The apex of a white slipper shell (*C. plana,* right) is more flattened than that of *C. fornicata.*

followed by degeneration of the male reproductive tract and development of female reproductive structures. Large older shells at the bottom of the stack are female, small younger ones near the top are male.

The white slipper shell (*Crepidula plana*), also known as the flat slipper shell, can be distinguished from *C. fornicata* by its size, only one inch in length, and its milky white color on both surfaces. In addition, the shell's apex is flattened and the internal shelf covers less than half the concave interior.

Both species of *Crepidula* are found attached to hard substrates throughout the study area.

Lunatia (Polinices) heros is the well known northern moon shell or northern moon snail. The snail is very common in shallow water with sandy or muddy bottoms, and if you follow a broad, short track that ends in a small mound you might find a moon shell under the mound. A broad, short track that has no mound at the end will probably yield a sand dollar. The thick shell of *L. heros* can be about four inches in diameter, usually with five whorls, somewhat compressed at the top; the aperture is large and oval with a horny operculum. Exterior shell color is grayish brown, with a purple-tinted center, and often has a yellowish brown periostracum. The interior color is grayish with some brown.

The moon shell uses its large, muscular foot to plow through the sand or mud in search of other mollusks, particularly bivalves (clams, etc.) (Color Plate 6b and 13c). This behavior leaves distinctive trails on the bottom. When it finds one, it uses its radula to drill a hole in the shell to get at the soft tissue within. Special glands near the mouth secrete an acid that weakens the bivalve's shell, facilitating drilling. The snail eats as many as four clams a day. It is common to find bivalve shells on the beach with holes, one eighth inch or more in diameter, having beveled sides; the signature of the moon shell. The oyster drill and sea star prey on the moon shell. The snail lays eggs, usually in late summer, and mixes them with sand, forming a characteristic "sand collar," so named because of the resemblance to Victorian period detachable shirt collars and the clergymen's collars (Figure 5-25). The collar shape is produced as the egg mass forms around the snail's foot. The egg masses occasionally wash ashore. Hold a sand collar up to the sunlight and the eggs can easily be seen. The large powerful foot, and the mucus a moon shell secretes, are very disruptive in a marine aquarium.

Figure 5-25. The "sand collar" of a moon snail. (Photo by Bradley Sheard.)

Neverita (*Polinices*) *duplicata* is commonly known as the lobed moon shell. The shell's spire is more compressed than that of *Lunatia heros* (Figure 5-26). A purplish brown callus (calcareous deposit) by the umbilicus (Figure 1-12), an indentation at the base of the columella, also distinguishes the snail. The dark center and eyeball shape of moon shells give rise to another common name, the shark eye.

Figure 5-26. Two views of the moon shells *Neverita duplicata* (left) and *Lunatia heros* (right). The spire of the lobed moon shell (*N. duplicata*), is more compressed than that of the northern moon shell (*L. heros*). The specimen of *N. duplicata* is 2.35 inches in diameter.

Figure 5-27. The lobed moon snail (*Neverita duplicata*) removed from its shell. Note the operculum attached to the muscular foot.

Whelks are the largest gastropods north of Cape Hatteras, with a shell reaching about eight inches in length and having a very large aperture. The shell is pear shaped; in southeast waters it is often called the pear conch. Around northern shores it is called whelk because whelk means "protuberance" and the animal's shell clearly demonstrates the meaning of the word. *Busycon canaliculatum* is commonly known as the channeled whelk. On the top, deep channel-like grooves are at the sutures of each whorl (Figure 5-29). Live specimens show a hairy skin-like periostracum. Exterior color is yellowish gray and the interior color is yellowish brown.

Native Americans used these large shells as drinking vessels and their sharp edges as cutting tools. Hoes were fashioned by jamming a pole thru chipped holes in the shells. Whelks are edible; the large muscular foot is used in scungilli and other dishes.

Whelks attach long, spiraling strands of flattened disc-like egg cases to rocks on the bottom, but heavy seas frequently break them loose, and occasionally a strand will wash ashore. Hold an egg case up to sunlight to see if young whelks are inside. When the parchment-like egg case is broken open, as many as 200 tiny miniature whelks are found inside.

Figure 5-28. The channeled whelk (*Busycon canaliculatum*).

The knobbed whelk (*Busycon carica*) is slightly larger than the channeled whelk, and can be easily distinguished by prominent knobs on the shoulders of the shell and the absence of channels (Figure 5-29). Even the egg strand can be distinguished by two sharp edges on each disc, while the channeled whelk's has only a single sharp edge (Figure 5-30).

Figure 5-29. The channeled whelk (*Busycon canaliculatum*, right) has deep channel-like grooves on the top. The knobbed whelk (*Busycon carica*, left) has prominent knobs on the shoulders. The channeled whelk is 3.5 inches across.

Figure 5-30. A 2-foot long egg strand of the knobbed whelk (*Busycon carica*) can be distinguished by two sharp edges on each disc (A), while the channeled whelk's, a 6-inch-long strand, has only a single sharp edge (B). Minute whelks removed from an egg strand disc (C).

A

B

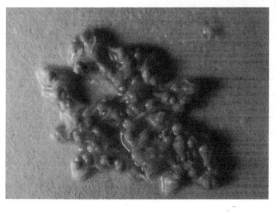

C

The channeled whelk and the knobbed whelk are both dextral (the opening is to the right). The lightning whelk (*Busycon contrarium*), found at Cape Hatteras and southward but seldom north, has small knobs on the shell, but is sinistral (opens to the left).

Whelks feed on other mollusks such as oysters and clams. They use their large, muscular foot to engulf the bivalve, then use their own shell edge as a wedge to force an opening through which they extend their proboscis and radula. Also, the radula can be used to bore a hole through the shell, and the soft body parts are sucked out.

What may appear to be a large, empty gastropod shell in shallow water may contain an octopus (*Octopus* sp.) (Figure 5-31). The animal is usually found in deep water, but on rare occasions a young one will be seen hiding in a subtidal crevice on a rock jetty or in an empty shell.

Ancient stories depicted the octopus as a monster that attacked ships and plucked terrified men from their decks like canapes. Actually, octopi are shy, intelligent animals with a strong parental instinct. The

Figure 5-31. An octopus may occasionally be found in shallow water.

illusion of the octopus as a monster was reinforced in 1866, when Victor Hugo stretched the truth in his novel, *Les Travailleur de la Mer* (The Toilers of the Sea). He recounts a vicious fight between a man and an octopus, ending with the man's death. His book set the standard by which the octopus would be described for nearly a century, and helped give the animal a common name, devil fish.

The octopus can easily be distinguished from squid by the absence of tentacles and fins. Its eight arms have two rows of suckers and are webbed for about one third their length. Due to the lack of fins and the shorter, thicker body, an octopus does not swim as gracefully as a squid. The octopus has an ink gland that secretes a brownish-black fluid that is stored in its ink sac. When the animal is alarmed, the ink is released through a duct located near the anus. The ink cloud provides the octopus with a means of escape from a predator.

The octopus feeds on crabs, lobsters, gastropods, bivalves, and fish if they can catch them. Also, they will eat each other. They eat at any time, but they forage away from home (a hole or cavity such as an empty gastropod shell) at night. The octopus kills its prey by a poison produced from the posterior salivary glands. When an octopus feeds on other mollusks it grasps its prey with its arms, and drills into the shell with the radula. Then, a viscous secretion is injected into the shell, weakening the gastropod so it can be pulled out of its shell. The radula looks like a parrot's beak; it is the hardest part of the animal's body.

Both sexes die after they have bred. The octopus is the most intelligent of invertebrate (no backbone) animals. Experiments have shown that an octopus can remember incidents and learn from experience.

The American oyster (*Crassostrea* [*Ostrea*] *virginica*), also called the common oyster, is an intertidal, filter-feeding bivalve mollusk; it cements one valve (one-half of its two-part shell) to a hard substrate, such as a rock. The thick shell varies greatly in size (to over eight inches) and shape, but is usually narrow and elongate. The two valves of most bivalves are equal in size, but in the sessile oysters the valve by which the mollusk attaches is smaller. The larval stage of the oyster attaches its mantle to the substrate with a drop of adhesive, then begins to secrete the shell. As the animal grows, the valve assumes the irregularities in the shape of the object it is attached to, producing a snug fit (Figure 5-32). Oysters have a large, centrally located, single adductor muscle. Exteri-

Figure 5-32. The shell of the American oyster (*Crassostrea virginica*) assumes the irregularities of the shape of the object it is attached to. Note the single adductor muscle scar. The specimen is 3 inches long.

or shell color is grayish; the interior is white with a purple adductor scar. Oysters can tolerate low salinity and are abundant in estuaries such as Chesapeake Bay. Pearls, when present, are not commercially valuable.

These familiar edible mollusks are considered by many to be the most commercially valuable invertebrate animal, and were once extremely abundant in the northeast. As filter feeders, oysters, like clams and mussels, may contain disease-causing organisms if collected in polluted waters. Type A hepatitis is an example of a disease acquired by humans eating tainted shellfish. Native Americans preferred them over clams; mounds of oyster shells (middens) are found at old camp-sites along the coast. Colonists considered clams the poor person's diet. Oysters were so prevalent on Long Island that they became the name-sake for Oyster Bay and Oysterponds. However, over-fishing, disease, and predators such as sea stars, whelks, and oyster drills have greatly reduced their numbers. A sea star may consume as many as six oysters per day. An oyster can produce half a billion eggs in a single season, but like other bivalve hatchlings few survive predation.

A small crab, *Pinnotheres ostreum,* is often found living with the oyster. The fact that females are found in oysters and the small size of their rounded carapaces accounts for their common name oyster crab (Figure 5-33). The male is smaller and free swimming and only enters the bivalve to mate with a female. It is a filter feeder, but occasionally feeds on the host bivalve. Some people consider them a delicacy and eat them raw, despite the "ticklin' of the throat" one may experience in the process. The small, about one-half-inch, carapace is whitish or pink in color. The crab is found throughout the study area. *Pinnotheres maculatum* is similar in appearance, but is found in mussels, scallops, and clams, and is also found throughout the study area. It is commonly referred to as the mussel crab.

Figure 5-33. The female oyster crab (*Pinnotheres ostreum*) is often found living within the valves of an oyster.

The gastropod *Urosalpinx cinerea,* commonly called the oyster drill, is found in oyster beds in this habitat and on rock beaches and jetties. The rather thick shell of this snail is rough, with a high spire. Revolving axial ribs are characteristic (Figure 5-34). The oval aperture's outer lip has several teeth. Shell length is approximately one inch. Exterior color is grayish and the interior is brownish purple.

As the common name implies, the snail feeds on oysters by using its radula to drill a hole through the bivalve's shell. The snail demineralizes the shell with an acid secreted by a gland in the anterior part of the foot. Once the outer part of the shell is softened, the radula is used as a drill for approximately one minute. Then the gland is applied to the site and acid is secreted for 30 to 40 minutes. The cycle is repeated until the

Figure 5-34. The oyster drill (*Urosalpinx cinerea*) uses its radula to drill a hole through a bivalve's shell, and is a scourge to commercial oyster beds. The specimen on the left is 1.2 inches long.

shell has been penetrated, taking several hours. The radula is then used to tear out small bits of tissue. The oyster drill is abundant throughout the study area and is a scourge to commercial oyster beds. However, an oyster can tolerate a lower salinity level than an oyster drill and the oyster larva that settles in brackish water (low salinity) may escape predation. Commercial oyster farmers often locate their oyster beds in areas where the salinity is less than 15 parts per thousand, to avoid the predators. The oyster drill is also found on pilings and rock jetties where it feeds on bivalves, barnacles, and other snails.

Most mollusks have clear or light blue blood. But *Anadara ovalis,* commonly called the blood ark, is among the few bivalves that have the red respiratory pigment hemogloblin. Thus, the blood gives the mantle and other tissues a bright red color. The many-ribbed shell is thick, and is approximately two inches long, with the umbos almost touching. The shell is white, with a protective brown covering called periostracum. The horny outer covering of organic material is secreted, like the shell itself, by the mantle.

Aequipecten (Pecten) irradians is a commercially important bivalve, commonly known as the bay scallop, and is found lying on the bottom in shallow water. Usually, at least in the United States, only the single large adductor muscle is eaten, and the remainder discarded. Each fan-shaped valve has prominent flattened radiating ribs that interlock at the rounded margins with the other valve. The valves have conspicuous wings on both sides of the umbo (Figure 5-35). By rapidly opening and closing its valves, the bay scallop can propel itself in convulsive jerks to escape predators. Scallops are one of the very few bivalves that neither attach to the substrate nor burrow into it. The rounded shell raises the animal above the bottom and draws in correspondingly cleaner water. The shell diameter can be approximately three inches. Exterior color varies considerably, ranging from gray-brown to brownish red. Colored rays are frequently present.

When the valves are open, two rows of small bright blue eyes can be seen on the mantle margin, just inside the shell. The eyes are complete with cornea, blue iris, lens, and retina. There is disagreement as to whether they can form images. Another common name is the blue-eyed scallop. Young scallops attach to eel grass until they are large enough to survive on the bottom. The bay scallop is found in large numbers from Cape Cod to Cape Hatteras.

Figure 5-35. Each fan-shaped valve of the bay scallop (*Aequipecten irradians*) has prominent flattened radiating ribs and conspicuous wings on both sides of the umbo. The valves are 3.25 inches across. Note the calcareous tubes of annelids on the valve on the left.

Rumors abound that some restaurants serve circular pieces of tissue cut from the fins of skates and rays as scallops. The muscle tissue of scallops separates easily, while that of a skate or ray will not.

Anomia simplex is commonly known as the jingle shell because a handful of the thin shells make a jingling sound when shaken. The shells are occasionally used in wind chimes. The upper valve is convex, rounded with an irregular margin, and yellow to orange in color. The lower valve is flat, rounded with an irregular margin, and yellowish white in color. The lower valve has an oval opening in the umbo region for the passage of a bundle of byssal threads for attachment (Figure 5-36). The shell can grow to two inches in diameter, but is usually one inch. Jingle shells are found attached to rocks, but the thin shells frequently wash ashore, although usually only the concave upper valve will be found. The empty shells of other bivalves are also found on the beach.

Just below the low-tide mark in shallow water many marine animals may be seen moving over the bottom. One of the more familiar and interesting to any beachcomber is the hermit crab (*Pagurus* spp). Its name derives from the fact that it lives in the empty shell of a gastro-

Figure 5-36. The lower valve of the jingle shell (*Anomia simplex*) is flat with an irregular margin and an oval opening in the umbo region for the passage of byssal threads. The upper valve on the right is 1 inch across.

pod mollusk. That behavior has resulted in the loss of the heavy abdominal exoskeleton that is characteristic of other crabs. The hermit crab is not a true crab. A true crab has a reduced abdomen that is folded under the body and cannot be seen from above, while that of a hermit crab is large and obvious (Figure 5-37). When threatened, the hermit crab withdraws into the cavelike protective shell and closes the opening with its pincer claws. However, some predatory fish swallow it shell and all. The hermit crab carries its portable house around with

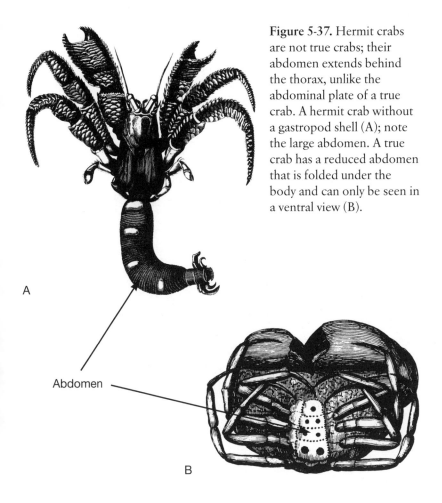

Figure 5-37. Hermit crabs are not true crabs; their abdomen extends behind the thorax, unlike the abdominal plate of a true crab. A hermit crab without a gastropod shell (A); note the large abdomen. A true crab has a reduced abdomen that is folded under the body and can only be seen in a ventral view (B).

A

Abdomen

B

it. Its large abdomen is twisted to fit around the central spire within the shell and the last pair of abdominal appendages are modified into hooks for clinging to the shell's spiral central support (Figure 5-37). The hermit crab holds so tightly to the shell that it will often be torn apart if forced out. The first pair of appendages have pincer claws, the second and third pair are used for walking. The hermit crab is a scavenger, but occasionally preys on small animals. Hermit crabs are usually very abundant in the study area. Male hermit crabs fight over females, or pull them around by the shell. The male is waiting for the female to shed her exoskeleton so he may deposit his sperm on her abdomen. She later uses it to fertilize her eggs.

Occasionally a hermit crab's gastropod shell will have an anemone attached to it. Both animals benefit from the association (symbiosis). The hermit crab carries the anemone along as it moves over the bottom and the anemone feeds on small pieces of food that float by as the hermit crab tears its prey apart. The stinging cells in the anemone's tentacles protect the hermit crab from predators. (Figure 5-38).

Figure 5-38. An anemone attached to the shell of a hermit crab will be moved to the new residence when the crab changes gastropod shells. The hermit crab uses its claws to stroke the anemone's column until it releases its hold on the shell. Then the crab places the anemone's base against the new shell until it adheres to it.

The common species are the broad-clawed hermit crab (*Pagurus polli-caris*) and the long-clawed hermit crab (*Pagurus longicarpus*). The former is larger, reddish or brownish in color, and adults often inhabit the empty shells of whelks (*Busycon*) or moon shells (*Lunatia* or *Neverita*) (Figure 5-39). *P. longicarpus* is smaller, and abundant in rock pools and shallow, sheltered waters. Adults are usually found in periwinkle (*Littorina*) shells or mud snail (*Ilyanassa*) shells. The main claws of the crabs may be used to distinguish between the two species; *P. longicarpus* (Figure 5-40) has a narrow claw while that of *P. pollicaris* is broad and flat. The left pincer claw is larger in both sexes. (Color Plates 12b, 12c, and 12d).

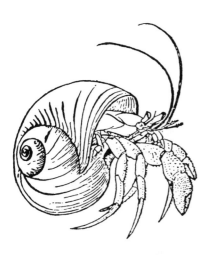

Figure 5-39. The broad-clawed hermit crab (*Pagurus pollicaris*) often inhabits the empty shell of the moon shell (*Neverita or Lunatia*). The pointed walking legs of the hermit crab allow the animal to drag its shell around.

Figure 5-40. A long-clawed hermit crab (*Pagurus longicarpus*) hitches a ride on the operculum of a channeled whelk (*Busycon canaliculatum*). (Photo by Mike Casalino.)

Hermit crabs are always willing to exchange their old home (shell) for a new one, and the selection process is amusing to watch. If you place several empty shells in an aquarium with a hermit crab, the crab will check them for size with its antennae. If one is a "good fit" it will quickly switch shells. Heating the shell with a match will usually force the hermit crab from its old home. However, if the hermit crab is stubborn remove the heat before it is harmed.

Hermit crabs in the study area, unlike the terrestrial species sold in pet stores, will die if removed from the aquatic environment. In Italy, hermit crabs are considered a delicacy and are cooked in oil and served in their mollusk shell.

A very small commensal colonial hydroid (*Hydractinia echinata*) is commonly called snail fur because it lives on the snail shells inhabited by certain hermit crabs. It is also found on rocks. The hydroid camouflages the hermit crab and, in return, receives a free ride and bits of food from the crab's meal (Color Plate 12b). With magnification, four types of polyps can be observed: a white nutritive polyp with tentacles, a club-shaped defensive polyp with many nematocysts, a sensory polyp that is similar in appearance to the defensive polyp, but contains nerve cells instead of stinging cells, and a reproductive polyp with reddish reproductive sacs. They are usually pink in color and are abundant from Canada to North Carolina.

Probably the most familiar crab is the blue crab (*Callinectes sapidus*), so named because there is usually some blue coloration, particularly on its pincer claws. Usually the tips of its claws are red (Color Plate 12a). The large pincer claws are able to inflict a painful wound to an unwary beachcomber. There is a large sharp spine on the lateral angle of the dark green carapace; the undersurface is white. A red-black line down the back forecasts the crab is due to molt its shell.

The blue crab is very common in bays and estuaries. The last (posterior) pair of appendages are modified into flat paddles for swimming and can be used as shovels for digging into the bottom when threatened by predators, leaving only the head and powerful pincer claws exposed (Figure 5-41). If a claw is lost, a new one will regenerate in about a month. The process of releasing a claw to a predator is called *autotomy,* which is further described in Chapter 6. The animal is a fast swimmer and feeds on small animals it easily catches, even fish. The large

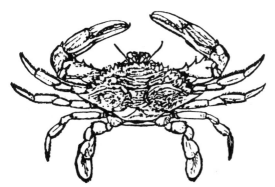

Figure 5-41. The flattened back pair of appendages of the blue crab (*Callinectes sapidus*) are used as paddles for swimming. There is a large, sharp spine on each side of the carapace.

crab, with a carapace that can be nine inches across, is highly valued for its edible qualities. It is referred to as a "soft-shell" crab after shedding its hard exoskeleton, and before a new one hardens, about 48 hours. The crab eats the molted shell to gain calcium for the developing new shell. A blue crab will molt 20 to 30 times during its three- to four-year lifetime. Females shed slightly fewer times than males.

In most species of true crabs the sex of the animal can usually be determined by looking at the ventral surface. A male blue crab can be distinguished by the inverted "T" shaped abdominal flap as compared to the much more rounded abdominal flap of a female crab (Figure 5-42).

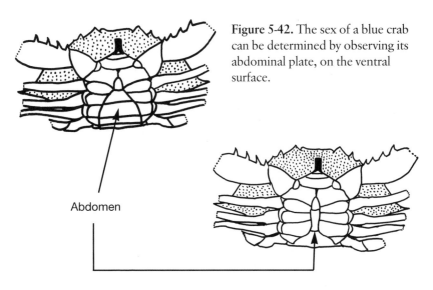

Figure 5-42. The sex of a blue crab can be determined by observing its abdominal plate, on the ventral surface.

Abdomen

The wider abdomen of the female facilitates carrying eggs. An immature female has a triangular abdominal flap. A female mates only once, but it lasts for 6 to 12 hours. Mating occurs in the bay and the female will produce about a million eggs, which she will carry to the ocean to release, but few hatchlings survive predation (Figure 5-43). It is estimated, that only one-half of one percent will live beyond one year. The surviving larvae are swept through inlets into bays (Figure 5-44). Young crabs hide in the eel grass. During winter months blue crabs move into deeper water. When handled or disturbed out of the water crabs produce a bubbly froth. Their gill chambers rapidly pump air instead of water, producing bubbles of air trapped in a thick mucus fluid.

Children can fish for crabs by tying a chicken neck to a piece of twine. Lower the bait to the bottom of the bay and after a few minutes

Figure 5-43. A female blue crab in sponge (with eggs). The female's abdominal plate is wider to accommodate the million or so eggs she will produce.

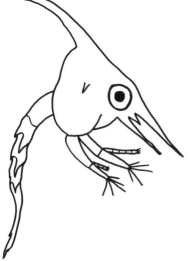

Figure 5-44. A larval stage of a blue crab. Most crustaceans undergo metamorphosis, a distinct change in body form, during their early development.

slowly raise it to the surface. If a crab is holding onto the bait, slip a dip net under it. Watch out for the claws; they can inflict a painful wound. Return females with eggs; most states have regulations setting size limits for blue crabs.

The mud crab (*Panopeus* and related genera) is easily recognized by the brownish color and black tips (fingers) of the large, unequal in size, pincer claws. It is usually found on muddy bottoms and feeds primarily on barnacles and bivalves. The large pincer claw can easily crush the shells of small clams and oysters, and human fingers if handled carelessly. The crab is also found on pilings and under rocks.

The stone crab (*Menippi mercenaria*), found from Cape Hatteras southward, seldom north, is a member of the same family and is similar in coloration. The claw meat is considered a delicacy and the collection of the crabs is regulated in most states (Figure 5-45). Some states allow only the larger claw to be taken and the live crab returned to the water where a new claw will regenerate in about two months.

The largest "true crab" and the most abundant in the bay habitat, also found subtidal off the ocean beach, is the spider crab (*Libinia emarginata*). Its long legs and sac-like body give rise to the common name (Figures 5-46 and 6-4). The broadly triangular, brown colored shell can be 4 inches and the leg span over 12 inches. In the Orient, the leg span

Figure 5-45. The claw meat of the stone crab (*Menippi mercenaria*) is considered a delicacy. The carapace of this specimen is .8 inch across.

of the giant spider crab can be 12 feet. The carapace is blanketed with hair-like bristles, spines, and knobs. Using a sticky mucus produced in its mouth, the crab sticks algae and other debris on the carapace for camouflage (Color Plate 11d). A spider crab cannot swim, so it walks along the bottom with its long legs. The crab is primarily a detritus feeder and the pincer claws are small and cannot pinch effectively.

Another familiar and unmistakable animal is the horseshoe crab or king crab (*Limulus polyphemus*). This only surviving member of a large, extinct group of animals is more closely related to spiders than crabs. Horseshoe crabs inhabited the seas 200 million years ago, long before *Tyrannosaurus rex* ruled the earth, and is often called a living fossil.

The body of a horseshoe crab consists of a large rounded, horse-shoe-shaped carapace and a long unjointed, spike-like tail (telson) (Figure 5-47). The lethal-looking telson is harmless. In the water the telson aids in maneuvering, and if the animal is overturned, it serves as a lever to right it. There are two large, conspicuous, compound eyes widely spaced near the middle of the upper surface of the cara-pace. Two small, simple eyes are located close together on the mid line near the front. Adult females can grow to about two feet long. They are most abundant along the shores of states with extensive wet-lands, which afford good breeding grounds.

The horseshoe crab is host to many other organisms, including bar-nacles, limpets, and algae, that attach to its hard shell.

Horseshoe crabs crawl along the bottom, plowing up the sediment with the rounded part of the carapace in search of the small burrowing animals on which they feed (Figure 5-48). They are often stepped on by

Figure 5-46. The long legs and sac-like body of *Libinia emarginata* give rise to the common name spider crab.

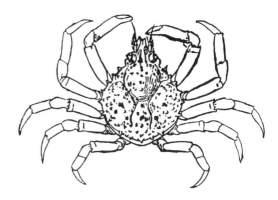

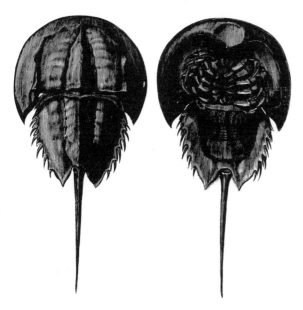

Figure 5-47. Dorsal and ventral view of the horseshoe crab (*Limulus polyphemus*).

bathers who, unfortunately, throw them up on the beach to die. Some fear them in the mistaken belief that the tail-like telson is filled with venom.

Despite their fearsome appearance, the much maligned horseshoe crabs are not only harmless; they are beneficial to humans. Scientists at Woods Hole Marine Biological Laboratory discovered in the mid-

Figure 5-48. A horseshoe crab (*Limulus polyphemus*) partially buried in the mud at low tide.

1960s that the horseshoe crab's blood clots when exposed to endotoxins (chemical poisons released by some bacteria). Thousands of horseshoe crabs are collected each year to donate blood for research. The blood is used to produce a white powder that detects bacterial contamination or disease. The horseshoe crabs are returned to the sea, unharmed by the experience.

The abdominal appendages of horseshoe crabs are modified for respiration. The first pair of broad, plate-like appendages covers the others, protecting them. The posterior five pairs are broad and thin, for gaseous exchanges. The overlapping, leaf-like appearance of these appendages gives them the name gill-books. They can be observed by turning the animal over, and gently raising the first pair of operculum-like appendages.

Horseshoe crabs mate in the spring or early summer, high in the intertidal zone at high tide. The larger female lays her eggs in the sand while the male clings to her back and covers them with sperm. As the tide ebbs, she covers them with sand; the eggs that survive predation by shore birds, crabs, and other animals hatch in about two weeks. The hatchlings look like miniature adults, minus the spiked tail.

The casts (molted shells) of horseshoe crabs are occasionally washed ashore. The old shell is split in the front, between the upper and lower carapace. The animal emerged through this opening, unlike true crabs that molt from the rear (Color Plate 11a).

Other creatures you may see near the water's edge include small fish, such as the silversides minnow (*Menidia*), easily recognized by the silver streak along the side.

The sea horse (*Hippocampus* sp.) is well camouflaged, but you may catch a glimpse of one hiding among the slender blades of eel grass in the shallow water of bays. The sea horse gets its common name from the horselike appearance of its head. It swims slowly through the water in an erect position, exhibiting no body movement except for a fanning motion of the fin on its back. The fish uses a prehensile tail, like a monkey, to hold on to a blade of eel grass (Figure 5-49). It has a long tube-like mouth through which it sucks up its prey, usually small crustaceans. The male has a pouch in the abdomen in which the fertilized eggs hatch and the young remain until they can survive on their own. Instead of scales, it has interlocking bony plates of body armor. What a wonderful creature!

Figure 5-49. The sea horse (*Hippocampus* sp.) is so named because of the horselike appearance of its head.

The pipefish (*Syngathus* species) has a long snout and breeding habits similar to the sea horse (Figure 5-50). The sea horse is strictly marine but the pipefish travels freely between salt water and fresh water. It is frequently found in eel grass beds. The pipefish has an unmistakable long, narrow body and large central dorsal fin.

The American eel (*Anguilla rostrata*) frequents shallow water. It is easily identified because its dorsal and anal fins connect to its tail fin (Color Plate 15c). The American eel is sometimes confused with the

Figure 5-50. The pipefish (*Syngathus* sp.) has a long snout similar to that of a sea horse and can live in fresh and salt water.

conger eel (*Conger oceanicus*), which usually prefers deeper water. The dorsal fin of the conger eel extends almost to the head (Figure 5-51), while that of the American eel extends only about three-quarters the length of the body. Conger eels do not enter fresh water, but the Americans eels do. Conger eels are found on the European and American sides of the Atlantic. American eels are rarely over 4 feet long, but conger eels can be over six feet. The largest conger eel caught was an eight-footer that weighed 128 pounds.

In late summer the shallow and warmer inshore waters often harbor tropical fish; they may be seen under docks, or by wading.

Figure 5-51. The dorsal fin of the conger eel (*Conger oceanicus*) extends almost to the head. Conger eels are found on both sides of the Atlantic.

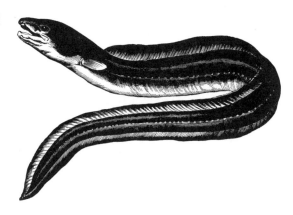

Bottom-dwelling fish such as skates and rays are occasionally seen in the shallow waters of bays and sounds. The fish have flat bodies and two enormous pectoral fins, sometimes called "wings." Although they are bottom dwellers, they are fairly good swimmers, flapping pectoral fins the way a bird flaps wings. Bottom dwellers do not get oxygen the way other fish do, which involves water entering through the mouth and exiting through the gills. Using that method they would ingest mud and other debris along with the water; instead water is brought in through spiracles (openings on top of the head by the eyes), and then passed through the gills (Figure 5-52).

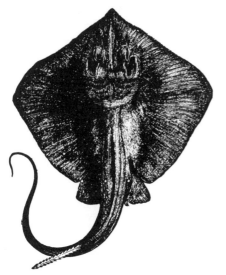

Figure 5-52. A sting ray. Note the spiracles behind the eyes.

The sting ray commonly found in the study area is *Dasyatus sabina.* The ray usually lies on the bottom and flutters its wings to cover itself with sediment, leaving only its eyes and spiracles showing. The sting ray has a spine in its snakelike "tail," which it wields in whiplash fashion. Its barbed stinger is used as a purely defensive weapon; if the animal is stepped on, the stinger quickly strikes. The stinger produces a jagged cut that is prone to infection. Also, the stinger is covered with a sheath of tissue that tears and releases venom. Pieces of sheath in the wound continue to release venom. The victim should remain quiet (activity increases circulation of the venom), flush the wound with clean, warm, water, and seek medical attention.

The primary diet of the sting ray is sand dollars, mollusks, worms, and shrimp that are crushed between strong, flat teeth. Sting ray embryos are nourished by a yolk sac, but are carried by the mother and are only released when they can fend for themselves.

The skate's (*Raja* sp.) tail is shorter than a ray's and lacks the stinger. The fish feed primarily on invertebrates that they find on the bottom. *R. eglanteria* is a common species in the study area; it can be identified by a ridge of thornlike spines on its back. Coloration is brown with darker brown spots on top, and white underneath (Color Plate 15b).

The common skate lays two egg cases per year. One surface of the black, leatherlike, rectangular egg case is usually covered with a sticky substance that gathers shells and stones to keep it from floating away. Also, the curly tendrils at each of the four corners often become entangled in eel grass and seaweed, helping to anchor it. Frequently, egg cases wash ashore after the embryo has hatched (Figure 2-30). Baby skates incubate within the case, often called a mermaid's purse, for 5 to 15 months. The yolk provides nourishment, and water passes through the egg case for gaseous exchange.

MARINE PLANTS

Seaweeds

Seaweeds (algae) can be found attached to substrates such as rocks and shells. However, there will be few, compared to the number in a rocky habitat (see Chapter 3). The few plants that have managed to attach themselves support even fewer animal grazers. Some loose mats of seaweed may be found washed up on the beach.

Green Algae. These algae have a characteristic grass-green color, and are well represented in the quiet water of bays and estuaries. Examples follow.

Ulothrix flacca has slender, slippery, unbranched filaments less than an inch long. The basal cell is somewhat modified to form a holdfast. A band-like chloroplast, which requires magnification to see, encircles the cell. *Ulothrix* thrives on rocks in quiet waters from Virginia northward.

Ulva lactuca is a bright-green membranous form, and is one of the most common green algae. The flat or ruffled membrane is two cells thick, with a rubbery texture. The plant is commonly referred to as sea lettuce because of its lettuce-like appearance (Figure 5-53). *Ulva* survives in a wide range of salinity, and is well adapted to brackish (mixed fresh and salt water) and moderately polluted waters.

Ulva grows to almost three feet in length, with almost no stipe at the point of attachment. The plant is usually found in the intertidal zone in less turbulent waters. When it forms large thin sheets of tissue, it often breaks loose and is seen floating in the bays during summer months.

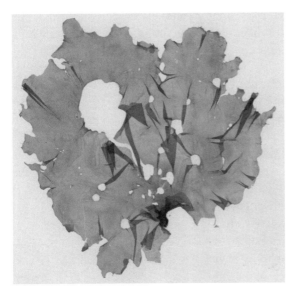

Figure 5-53. The green alga *Ulva lactuca* is commonly referred to as sea lettuce because of its lettuce-like appearance. The specimen is 5.5 inches across.

Ulva is notorious for clinging like a piece of cellophane to the bodies of bathers. Holes in the blade are due to feeding of grazing animals. Sea lettuce is edible; it can be cut into narrow strips and added to salads. Ducks eat *Ulva* and *Enteromorpha*, another green alga.

Monostroma species can easily be confused with *Ulva*. *Monostroma* consists of a single sheet of cells, making it less coarse than *Ulva*. These algae, which closely resemble each other, can be positively identified only by microscopic examination of their arrangement of cells. A rapid, but less certain means of identification can be made by touch. *Ulva* feels sturdier than *Monostroma* because it is thicker. Fingertips can usually be seen clearly through a single layer of *Monostroma*—but not through *Ulva*. *Monostroma* grows to about eight inches in length.

Members of the genus *Enteromorpha* are distinguished by their tubular body form or their expanded blade with a tubular base. The species with an expanded blade are often confused with *Ulva* and *Monostroma*. They also tolerate a wide range of salinity and are found in the intertidal zone in less turbulent waters. They are thin, ribbon-like algae that are tolerant of varying environmental conditions such as salinity and pollution. They are often called sea hair and are most abundant in salt marshes and bays (Color Plate 3a). They are also found on jetties and in tidal pools. Most are exceedingly variable in form; some are notoriously difficult to identify. Observation of chloroplasts with a microscope is often required.

Enteromorpha linza has a solid unbranched, flattened blade, except along the margins, which are frequently hollow and wavy. The ribbon-like blade is usually spirally twisted. Size ranges up to more than 16 inches in length and over one-inch in width. The width in proportion to length is highly variable (Figure 5-54).

Enteromorpha intestinalis is unbranched and, as the specific name implies, is tubular, long and sometimes filled with gas (like the intestines of humans) (Figure 5-55). It is about ¼-inch wide and may be more than 12 inches in length. The tubes can become inflated with gas during photosynthesis, giving them an irregularly puffed out shape. Chloroplasts are cup-shaped. This species will thrive in polluted waters.

Enteromorpha compressa is distinguished by small side branches off the main tube; both are usually flattened or compressed. Chloroplasts are cup-shaped and contain only one pyrenoid. It is usually less than 12 inches in length.

Figure 5-54. The green alga *Enteromorpha linza* has a solid, unbranched, flattened blade. The specimen on the left is 7.5 inches long and about 1 inch wide.

Figure 5-55. *Enteromorpha intestinalis,* like the specific name implies, is tubular, long and sometimes filled with gas. The specimen is almost 7 inches long, and .2 inch wide.

Enteromorpha flexuosa also has side branches off the main tube. It is usually less than 12 inches in length. Chloroplasts are star-shaped, and have two or three pyrenoids.

Chaetomorpha linum is an unbranched, coarse and wiry mass of hair-like filaments (Figure 5-56). Its thick cell walls are easily observed under the microscope. It often forms masses that give the appearance of tangled string. The plant has a yellow-green color.

Cladophora species are erect tufts of fine, highly branched, filamentous algae (Figure 5-57). The various species are usually distinguished by their cell size and branching pattern. If a specimen removed from the water collapses into a mass of green, with the branches losing identity, it is *C. crystallina.* If it goes limp but doesn't collapse completely, it is *C. gracilis.* Both species can be found throughout the intertidal zone, attached to jetties and pilings in the spring and summer, but free-floating for most of their existence. They form large masses on the surface of quiet bays.

Codium fragile is composed of compact, interwoven, filaments that form large, dichotomous (forking) branching, solid cylinders ¼-inch in diameter. It has a sponge-like or velvety texture and is commonly

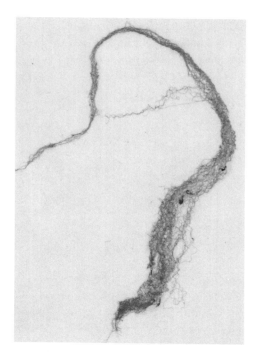

Figure 5-56. The green alga *Chaetomorpha linum* is an unbranched, coarse and wiry mass of hair-like filaments. The specimen is about 8 inches long.

Figure 5-57. *Cladophora crystallina* is a fine, highly branched, filamentous green alga. The specimen is 4.5 inches long.

known as green sponge seaweed (Figure 5-58). The Greek word codium means "skin of an animal," referring to the plant's soft, spongy texture. The plant may be 16 inches long, and equally broad. When the alga is pulled apart, the broken ends seal themselves immediately to prevent water loss. It grows on shellfish and rocks, usually in quiet bays, in the low intertidal zone. *Codium fragile* was non-existent on this side of the Atlantic prior to 1957. In that year, a few plants were found at Orient Point, Long Island, New York. It is speculated that *Codium* was unintentionally brought over on the hull of a ship. Regardless of how it was transported, it has flourished since its introduction. It has found an ideal niche, has no natural predators, and has spread throughout the study area to become one of the most common seaweeds.

Brown Algae. These algae have a characteristic brownish color; an example is *Ectocarpus* spp., a small filamentous branched, yellow-brown plant, generally an inch or two long (Figure 5-59). The filaments are uniserate (a single row of cells). It is usually epiphytic on other algae or eel grass, but is also found attached to rocks and pilings.

Figure 5-58. The green alga *Codium fragile* has a sponge-like texture and is commonly known as green sponge seaweed. The specimen is 5.75 inches long.

Figure 5-59. Four specimens of *Ectocarpus* spp., a small filamentous brown alga. The specimen on the right is 2 inches across.

Red Algae. Species of red algae are occasionally found in bays and have a characteristic red color. Examples follow.

Agardhiella tenera has relatively slender alternate (one branch at a node), bright red, translucent branches that narrow somewhat toward the main axis and gradually taper out to the apex (Figure 5-60). The plant is large, often growing to 12 inches, and is usually bushy.

Gracilaria spp. is similar in appearance to *Agardhiella,* but is a paler, purple to olive-green color, and somewhat smaller (Figure 5-61). While the branches of *Agardhiella* are round and tapered at the base, those of *Gracilaria* are flattened at the base. It is frequently found floating in large masses above the bottom. *Gracilaria* is an important source of agar, which is used in hand lotions, to process film, and for bacterial cultivation.

Dasya pedicellata is one of the most beautiful red algae; it makes a very attractive mounted specimen. It is easily identified by the delicate red, uniseriate hairs, up to ¼-inch long, covering the cylindrical main axis and branches, giving each branch a featherlike appearance (Figure 5-62). The plant is often referred to as chenille weed and can grow to two feet in length.

Figure 5-60.
Agardhiella tenera has slender, bright red branches that narrow somewhat toward the main axis and gradually taper out to the apex. The specimen is 7.75 inches long.

Figure 5-61. The red alga *Gracilaria* sp. is similar in appearance to *Agardhiella,* but is purple to olive-green in color, and is somewhat smaller. The specimen is 3 inches long. The slightly enlarged, darker areas on the branches are reproductive structures.

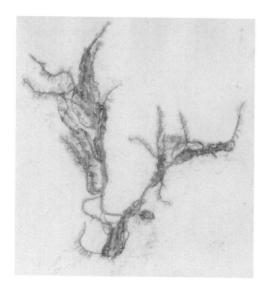

Figure 5-62. *Dasya pedicellata* has delicate red hairs covering the cylindrical main axis and branches, giving each a featherlike appearance. The specimen is 4 inches long.

A red alga that often appears brown in color is *Spermothamnion turneri.* It grows in spherical shaped tufts attached to eel grass or other algae. Strong currents or waves tear them loose and they wash ashore. Under the microscopic you can detect the lateral filaments branching out from the main axis in a pinnate fashion, which identifies the alga.

Flowering Plants

These plants are represented by *Zostera marina,* which is not a member of the grass family, but is commonly known as eel grass. It is easily identified by its erect shoots and underground, stemlike rhizome, which anchors the plant in the soft bottom (Figure 5-63). The shoot consists of a very short stalk, which bears ribbonlike leaves that are long and thin, resembling blades of grass, and may reach three feet or more in length. The leaves are lost in the fall, and wash up on shore in large masses. Tiny green flowers develop in grooves on a leaflike spike.

Eel grass is common to the bays and estuaries of the study area where there is little water motion. It is rarely completely exposed at low tide but its long, flattened, bootlacelike leaves can be seen waving on the surface of the water, giving the appearance of submerged meadows.

Eel grass beds support a very active animal community, some living directly on the plants themselves. *Zostera* is often covered with snails,

Figure 5-63. A young 6 inch long specimen of eel grass (*Zostera marina*). Note the stemlike rhizome and roots.

hydroids, small crustaceans, and tube worms. Colonies of *Pennaria tiarella,* an alternately branching, yellow-to-black in color, hydroid that grows to about six inches, is frequently found on eel grass, seaweeds, and hard substrates such as pilings. The polyps of this animal are flask-shaped. Seaweeds will also use the grass as a substrate to live on, and snails feed on the algae. *Thais,* the dog whelk, preys upon the other snails and in turn is eaten by the hermit crab. The hermit crab then takes the empty snail shell, if it "fits," as its home. Gelatinous masses of bright yellow snail eggs are occasionally found attached to eel grass.

Haliclona loosanoffi, a sponge, is frequently found attached to eel grass, algae, or the undersurface of rocks in the lower intertidal zone. The color ranges from yellow to reddish-brown. Oscula, about one-inch high, are on raised branches of this encrusting sponge. During the summer, cold-resistant gemmules (food-laden cells covered by a firm coating of spicules) are formed at the base of the animal. By fall, the parent sponge disintegrates, leaving only the gemmules. In the spring, the gemmules develop into new sponges, which can grow to about three inches in diameter.

Eel grass almost disappeared in 1931–32 because of a parasitic aquatic fungus. Its decrease reduced the population of animals such as the delicious bay scallop that depended on eel grass beds for shelter and food. The plants developed a resistance to the fungus, but it took almost 15 years for *Zostera* to return to its pre-1931 growth. Most of the marine debris at the high tide mark on many bay beaches, especially in late fall, will consist of dead leaves of eel grass.

ESTUARIES

An estuary is the widening lower reaches of a river system that is influenced by tidal waters. It includes a bay that extends inshore to merge with the river water. Waters of the sea rise and fall with the tides; those of the rivers are always moving seaward. Estuarine water is a unique mix of fresh and salt water known as brackish water, and is of constantly varying salinities.

Marine organisms in estuaries are limited to those that can adapt to changing salinity of the water. Anemones, blue crabs, a few snails, some bivalve mollusks, and annelid worms are among those that can tolerate the rapid changes of salinity.

An excellent way to view bottom dwellers in bays and estuaries is to use a face mask, snorkel, and swim fins and slowly swim or drift in shallow water. But be wary of boat traffic and currents that might be difficult to swim against. To prevent fogging of the glass plate in the face mask, spit on the inside glass surface, then rub it around, and rinse with clear water.

The apparently barren expanse of exposed sediment has been shown to contain many burrowing animals, so look beneath the surface as well as on it. A messy but effective way to view burrowing animals at low tide on exposed mudflats is to dig a trench a foot deep, six inches wide and three feet long. Carefully excavate sediment along the sides of the trench to expose tube worms and other organisms. When finished, recover the animals. Hip waders or shorts will reduce the laundry load.

6

Salt Marshes

*To a person uninstructed in natural
history, his country or seaside stroll
is a walk through a gallery filled with
wonderful works of art, nine-tenths of
which have their faces turned to the wall.*

Thomas Henry Huxley

Salt marshes are low-lying areas of sand and mud that are covered with fields of grasses belonging mostly to the genus *Spartina*; they border estuaries and bays protected from the open ocean. The wetlands are interwoven with meandering tidal creeks and crisscrossed with mosquito control ditches (Color Plate 1b). The marshes are submerged and flushed twice each day (actually every 24 hours and 50 minutes) by the tides.

Salt marshes are present throughout the study area—on Cape Cod at Barnstable and on the lower Cape, and scattered along the southern New England coast from southern Massachusetts through Rhode Island and Connecticut. Marshes are found around Peconic Bay on Long Island and occur on the western half of the south shore. The western-most of Long Island's marshes is in Jamaica Bay, within the boundaries of New York City. Marshes are also found on Staten Island. The New York City marshes are heavily polluted and the least attractive to visit.

South of New York, salt marshes are more extensive. In the 1970s, New Jersey had about 350 square miles of salt marsh. The Cape May marshes harbor millions of shore birds and waterfowl during the migration season.

The western side of Delaware Bay has extensive salt marshes. They are abundant along the eastern shore of Chesapeake Bay, but less frequent on the western shore. Salt marshes are most abundant along the shores of Albemarle Sound in northeast North Carolina.

MARSH PLANTS

There are approximately 6,000 species of grasses in the northeastern United States. Those we are more familiar with, the species that are present in our lawns, are relatively short plants. However, the predominant and conspicuous ones in the marsh are taller specimens.

The tall grass along marsh creek banks and in the lower areas of a marsh is *Spartina alterniflora* (commonly known as salt marsh grass or cordgrass). The plant grows to about five feet in height and the lower part of the plant is covered with water at high tide (Figure 6-1). Green, inconspicuous flowers bloom in September and October. The colonists

Figure 6-1. Salt marsh grass (*Spartina alterniflora*). The seed head on the right is 10.5 inches long.

used *S. alterniflora* as thatch for the roofs of their cottages. Brown algae and green algae will be found growing at the bases of salt marsh grass (Color Plate 3a).

Species of the brown alga *Fucus*, found here, are conspicuously different in appearance from the specimens observed on exposed rock jetties. Transplantation experiments indicate that some of the morphological differences result from the ecological setting rather than genetic factors. The roots of *Spartina* stabilize the sediment much as bay sediments are stabilized by eel grass.

At the high tide level *Spartina patens* takes over. The plant is often called salt meadow grass or salt hay. The grass was once cut for hay, usually in August, as feed for farm animals during the winter; that area of the marsh was referred to as the salt meadow, thus the plant's common names. Salt meadow grass is considerably shorter than salt marsh grass and grows to about two feet in height (Figure 6-2). Flowers that

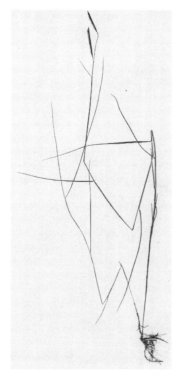

Figure 6-2. The lower seed head of this specimen of salt meadow grass (*Spartina patens*) is 1.25 inches long.

are green and inconspicuous bloom from late July to October. Salt meadow grass grows in stands and marks the average level of the marsh.

S. alterniflora washes away in winter, leaving only a stubble, while *S. patens* turns brown in color, but remains throughout the winter (Figures 6-3 and 6-4). In early summer the pale green shoots of young plants can be seen projecting out of the mud. Both species of *Spartina* are found from Newfoundland to Florida.

Figure 6-3. During the winter salt marsh grass (*Spartina alterniflora*) washes away and only stubble remains (area nearest water in the photo), while salt meadow grass (*S. patens*) remains throughout the winter.

Figure 6-4. In early spring, a dead spider crab (*Libinia emarginata*) is surrounded by dead salt meadow grass (*Spartina patens*).

The salt meadow area of the marsh is submerged only during extreme high tides and storms. Stands of spike grass (*Distichlis spicata*) will be found mixed in with salt meadow grass, although not as abundant. Spike grass is similar in appearance to salt meadow grass except when in flower (Figure 6-5). The fruiting (reproductive) structure is different, but the flowers are similar and bloom from August through October. Salt meadow grass often grows in mats that give the appearance of "cow licks," like a head of uncombed hair. Spike grass is found from Prince Edward Island to Florida. This area of the marsh is more productive than any other natural community of plants. About 2.5 acres can produce 10 tons, dry weight, per year of plant material, four times as much plant growth as a cornfield.

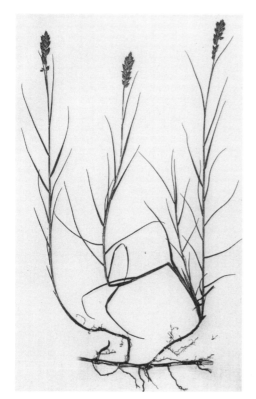

Figure 6-5. Spike grass (*Distichlis spicata*) is often mixed in with salt meadow grass. The specimen on the left is 10.5 inches long. Note the rhizome, a horizontal underground stem.

The most conspicuous plant, *Phragmites communis* (reed grass), is found in fresh and brackish water from Nova Scotia to Florida. The reed is as tall as a human, and occasionally twice as tall. The characteristic dense clusters are usually pure stands with nothing else growing within. Each stand contains plants that grow to about the same height, producing a level top. By August, the tops of the plants have highly-branched feathery plumes, in which purple-gray flowers are produced. When the plumes mature, they range in color from light to very dark gray; those of each stand will be the same shade of gray (Figure 6-6). The dense clusters of *Phragmites* may be found in small stands of a few square yards or large ones in stands of more than an acre.

All grasses are flowering plants, but the ones mentioned above, once established as seedlings, propagate by rhizomes. A rhizome is a horizontal underground stem with new plant shoots extending up from the upper surface and roots emerging from the undersurface (Figure 6-5).

Figure 6-6. An 11.5-inch-long highly branched feathery seed plume of *Phragmites communis.*

Cattails may be found where fresh water flows onto the marsh from a creek or hidden spring. The narrow-leaved cattail (*Typha angustifolia*) can be distinguished from the common cattail (*Typha latifolia*) by the short interspace between the brown, sausagelike pistillate (female) and paler staminate (male) portions of the flowering spike. The two flowering structures touch in *T. latifolia* (Figure 6-7). Both species are in bloom from late May through July, and are found from Maine to Florida. The lower pistillate portion remains through the winter (Figure 6-8). The bladelike leaves of the narrow-leaved cattail are narrower, as the name implies, than the common cattail. The common cattail grows to three to nine feet tall while the narrow-leaved cattail is usually two to five feet tall.

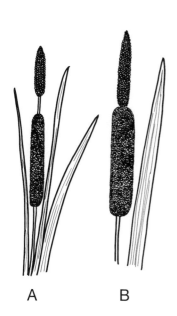

Figure 6-7. The narrow-leaved cattail (*Typha angustifolia*) (A) can be distinguished from the common cattail (*Typha latifolia*) (B) by the interspace between the lower pistillate and upper staminate portions of the flowering spike. The two flowering structures touch in T. latifolia.

A   B

The glasswort (*Salicornia* species) thrives in low areas in the marsh where saltwater accumulates, then is concentrated by evaporation. The common name originated from the crunching sound the plants make underfoot. The ankle-high, odd-looking plant has red or green stems, resembling fingers poking up through the mud, and lacks obvious leaves; the leaves are reduced to scales along the jointed stem. The glasswort looks like long, jointed green pipe-cleaners (Figures 6-9 and 6-10). The green flowers are inconspicuous and bloom from August

Figure 6-8. Table tops of ice, resting on the stubble of salt marsh grass (*Spartina alterniflora*), outline the meandering tidal creeks of a salt marsh in February, Southampton, Long Island. Note the cattails (*Typha* sp.) in the foreground and reed grass (*Phragmites communis*) in the left background.

Figure 6-9. The specimen of glasswort (*Salicornia virginica*) is 10.5 inches across.

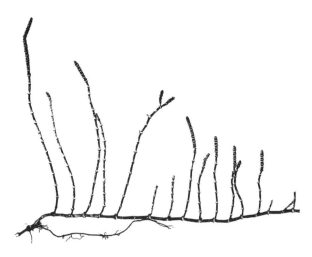

Figure 6-10. From the bottom of the root to the top of the plant, the specimen of glasswort (*Salicornia bigelovii*) is 5.25 inches in height.

through November. It is reported that it can be pickled or added to salads in which its briny taste adds a bit of spice. *Salicornia* species are found from Nova Scotia to Georgia.

A marsh plant with small but obvious yellow and lavender-colored flowers is the sea lavender (*Limonium* species). The low shrub grows to two feet and has large lancelike leaves at the base, with small stems extending upward. The flowers bloom from July through October. Sea lavender is found from Newfoundland to Florida.

The terrestrial plants that dominate the marsh vegetation have adapted well to an environment where they are at least periodically submerged in saltwater, which they convert to fresh water by concentrating salts in their root cells. Salt glands in the surface tissues remove excess salt from the internal fluids and deposit it through pores to the outside of the plant. Salt crystals, sparkling in the sun, can usually be observed on *S. alterniflora* leaves and stems.

MARSH ANIMALS

A variety of animals, such as birds, rabbits and deer feed on *Spartina* species. Also, at the end of the growing season, the decomposing marsh grasses provide the estuaries and bays with nutrient-rich detritus that is important as a food source for the substrate-ingesting and filter-feeding animals. The detritus is the start of a food chain leading to the fish that inhabit our coast. Receding tides carry decomposing marsh grasses (detritus) to the bays and estuaries, where many animals depend upon it for food. Detritus feeders (e.g., amphipods, isopods, worms) are eaten by predators such as crabs and small fish; the small predators are preyed upon by larger fish, ducks, and other animals. In turn, they are eaten by yet larger predators such as osprey and humans.

Although plants are the most abundant organisms of the marsh, these shallow, quiet waters are integrally related to shellfish and finfish productivity; they are breeding grounds for many types of marine animals. The wetlands are the ocean's nursery grounds, where most estuarine marine life develops into adults. Many species of commercially important fishes live here during their juvenile stages. Others such as bluefish, fluke, and flounder feed here. The marshes are among the most productive habitats in the world and swarm with shrimp, blue crabs, killifish, pipefish, sticklebacks, and eels, which in turn attract the herons and egrets, and a variety of other wading birds. Many of the organisms in habitats previously described (e.g., mussels, oysters, scallops, barnacles, mud snails, isopods, and amphipods) are found in the wetlands. The ribbed mussel will often be attached by byssal threads to the roots of *Spartina alterniflora*, with only an inch of the valves exposed above the mud.

Three species of periwinkle snails are common to the salt marsh. The common periwinkle (*Littorina littorea*), also frequently encountered on rock and other hard substrates, is usually the most numerous; it is found from Canada to New Jersey. The shell surface is smooth, the color is variable, and the length can be about one inch (Figure 6-11, and Color Plate 6c). The northern rough periwinkle (*Littorina saxatilis*), also known as the marsh periwinkle, is found from Labrador to New Jersey. The surface of the shell is rough, the color is variable, and it is usually less than ½-inch long. The northern rough periwinkle is cutting its ties

Figure 6-11. The common periwinkle (*Littorina littorea*). The specimen on the left is .9-inch-long.

with the sea and is becoming a land animal. It is viviparous (gives birth to live young) instead of depositing eggs on rocks in the water, like marine periwinkles and other marine snails. Also, unlike other periwinkles, it possesses a gill cavity that is supplied with blood vessels and, like a lung, breathes oxygen from the air. Constant submergence would be fatal to the snail. The southern periwinkle (*Littorina irrorata*), also called the Gulf periwinkle or lined periwinkle, is often seen on the stems of *Spartina*. The shell is more pointed than that of the common periwinkle and has spiral rows of dark spots. It is occasionally found on jetties and is found from from New York to Florida. The snail uses its radula to feed on the living plant and to scrape off the residue of plant detritus left behind by the ebbing tide. The radula has up to 300 rows of teeth and the rasping action wears out about five rows a day. More rows of teeth are produced to replace the old ones. As the tide rises, the snail migrates up the plant to escape waterborne predators such as the blue crab (*Callinectes sapidus*).

Periwinkles are found during the warmer months and are eaten by shore birds. The common periwinkle (*L. littorea*) is edible. After cleaning, it is steamed for about ten minutes, then the small snail can be removed from the shell with a straight pin and dipped in garlic butter.

Do not confuse periwinkles with the inedible mud snails (*Nassarius* species) described in Chapter 5.

The most common, and one of the more interesting inhabitants of the salt marsh, is the fiddler crab (*Uca* species). There can be more than a million fiddler crabs per acre, and if you move slowly through the marsh grass you might hear a rustling sound as they scurry away en masse. Both sexes of juvenile crabs have small claws. However, adult male crabs are easily recognized by their much enlarged claw, either right or left, but usually the right (Figures 6-12 and 6-13). The other claw is about the same size as those of the female. If the large claw is lost, the other will enlarge until it is the size of the original fiddle claw; the regenerated new claw will be the normal size. The male brandishes his large claw in the presence of a female, a posturing that is part of the courtship ritual, which occurs each month four to five days before spring tide. The horizontal movement of the large claw gives rise to the common name fiddler crab. The small (their squarish carapace is about one inch across) crabs live in burrows, usually in the intertidal zone, in the banks of marsh creeks. Their slanting tunnels are usually one to three feet deep and often end in a horizontal room. Dime-sized openings in

Figure 6-12. Male and female (on right) fiddler crabs (*Uca* sp.) The female's carapace is .65 inch across.

Figure 6-13. Male fiddler crabs (*Uca* sp.) have an enlarged claw, usually the right one. (Courtesy of the National Park Service.)

the mud, with stacks of mud pellets nearby indicate a newly excavated burrow (Figure 6-14). During excavation of the burrows, the crabs scrape mud and sand into pellets and bring the excavated material to the surface attached to their legs. Fiddler crabs are in tune with the rise and fall of the tides and plug the burrow opening with mud during high tide.

Figure 6-14. Entrances to fiddler crab (*Uca* sp.) burrows. Note the small, light colored balls of dirt in the lower corners of the photograph.

The crabs emerge from their burrows only during the ebbing tide to search for algae and detritus to eat. They are preyed upon by fish, other crabs, and shore birds, and are used as bait by fishermen. The crabs have reduced-size gills but possess a highly vascularized branchial (breathing) chamber that only needs to be damp to allow gaseous exchanges. The non-swimming fiddler crabs can remain out of the water for about a month. Shell coloration is affected by the solar cycle and the crabs are dark, blackish brown during daylight hours, but gradually become light gray in color as the sun begins to set. They are dormant during winter months, remaining in their burrows in a sleepy stupor.

Another crab common to the wetlands is the marsh crab (*Sesarma reticulatum*). It also lives in burrows that are in the intertidal zone or at the high tide line under seaweeds. Shell color variations may be black, olive, or purple. Slightly larger than the fiddler crab on which it occasionally preys, its primary food is algae and *Spartina* (Figure 6-15).

The green crab (*Carcinus maenas*) is easily recognized by the greenish color of the surface of the upper shell (Color Plate 13a). The surface of the lower shell of adult males is yellowish while that of an adult female is reddish-orange. The green crab is a swimmer, but its last pair of legs are only somewhat flattened, not paddle-shaped as in the blue crab. The crab grows to three inches. It is found intertidal, but is also common in other habitats, such as rock jetties, and is indifferent to the degree of salinity. There is no hardier animal on the shore. The French

Figure 6-15. Diagram of the carapace of the marsh crab (*Sesarma reticulatum*).

gave it the common name *le crab enragé,* because of its pugnacious habit of meeting intruders in an attack mode, with open pincer claws. Like most crabs, it is a voracious predator, feeding on mollusks, annelids, and other crustaceans. The green crab, in turn, is preyed upon by fish and shore birds. It is found from New Jersey northward. (Figure 6-16).

A

B

Figure 6-16. Frontal view (A) and back view (B) of a green crab (*Carcinus maenas*).

The green crab was a European species that was accidentally introduced to the United States on, or in, a wooden sailing ship. The crab was not observed north of Cape Cod during the 19th century, but today it is one of New England's most common intertidal crabs.

All crabs and related decapod crustaceans, such as the lobster, are capable of a process known as *autonomy,* or the throwing off of legs, particularly the pincer claws. This is the ability to shed a limb that is trapped by a dislodged rock or in the grasp of a predator. The separation occurs along a predestined breaking plane, an encircling groove near the base of the third limb segment, counting from the junction with the body. A special autotomizer muscle bends the limb at such an extreme angle that it breaks along this plane of weakness (Figure 6-17). Surprisingly, this is a reflex action, an automatic response, that the animal cannot control, an adaptation to ensure survival.

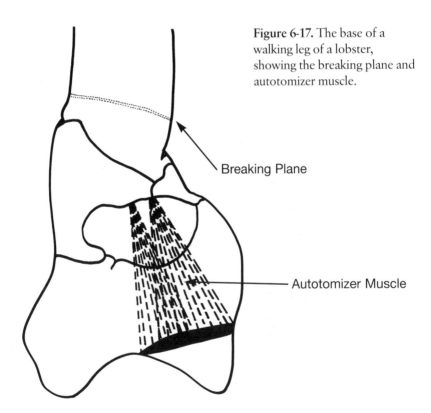

Figure 6-17. The base of a walking leg of a lobster, showing the breaking plane and autotomizer muscle.

Breaking Plane

Autotomizer Muscle

The depth of the encircling groove ensures that little tissue damage occurs and the small exposed area is covered with a thin membrane, perforated in the center where blood vessels and nerves are withdrawn. A blood clot closes the opening and superficial tissues that form the shell extend over the stump and form a small projection from which a new limb is eventually regenerated. After the next molt, a miniature limb will be present and increases in size with each successive molt until it reaches the original size. Occasionally a crab or lobster will be seen with a partially regenerated limb.

The glass shrimp (*Palaemonetes* species) is frequently seen swimming in the shallow water. The common name is derived from the animal's small, translucent body. Adults can be two inches in length, with small brown spots or streaks scattered over the body. Glass shrimp are also abundant in the shallow waters of bays and in beds of eel grass. Ducks, eels, and other animals feed on the small arthropod.

The sand shrimp (*Crangon vulgaris*) is also abundant in the marshes and bays of the study area. The body color varies from a pale gray, with small dark spots, to mottled brown. The shrimp blends into a sandy background and is difficult to detect when motionless. Pigment cells allow the animal to conform in color to different sediments. The sand shrimp is larger (up to three inches in length) than the glass shrimp, and the first pair of walking legs have large claws (Figure 6-18 and Color Plate 10c). The female lies on her side during copulation and eggs are attached

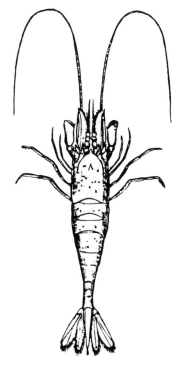

Figure 6-18. The sand shrimp (*Crangon vulgaris*) is abundant in the marshes and bays. Note the large claws on the first pair of walking legs.

to the base of the last two pairs of walking legs. Sand shrimp migrate into bays and estuaries during the summer, but move to deeper water in the winter. The sand shrimp is the common edible shrimp of Britain. Several species of fish also feed on them.

Shrimp feed on plants, eggs, and animals such as smaller crustaceans, mollusks, worms, and larval stages of fish. They flourish in brackish water and are common in estuaries. Shrimp can be caught at low tide by sweeping a long-handled net in front of you over the surface of the sand in shallow water.

When visiting the marsh during warm months, bring insect repellent. Voracious insects like mosquitoes, "no-see-ums" midges, greenhead flies, horseflies, and deerflies thrive and breed in the wetlands. In vicious blood lust, they feed on beachcombers and in turn provide food for the migratory birds.

Salt marshes smell! The natural pungency, its "essence 'd estuary," is produced by the decay of plants and animals. The decomposing organisms release nitrogen into the air and ammonia compounds to the mud. Like bay sediments, the marsh mud releases hydrogen sulfide (an odor reminiscent of rotten eggs) when disturbed.

Marsh music is produced in the backwoods by male birds, tree frogs, and crickets.

One of the best and least messy ways to see the marsh is to use a canoe or kayak (at high tide) and slowly and silently paddle up the tidal creeks.

SAVE THE MARSH

Not everyone appreciates the vital role of the marshes in the ecological balance of the seas. Many consider the wetlands as foul smelling, muck-ridden places to be avoided. However, the decomposing marsh grasses produce a valuable detritus, nature's soup of nourishment and energy for organisms in the salt marsh, bay, and ocean. Detritus fertilizes the mud when decomposer bacteria fix nitrogen in surface layers, making this valuable nutrient available for continued plant growth. Receding tides also carry detritus to filter-feeding animals and animals that feed directly on the mud. The small animals are in turn preyed upon by larger animals such as fish and birds.

With mankind's desire to live by the water, wetlands have suffered from filling and development. The United States has lost more than half of its wild and beautiful marshes in the past 200 years. New York has lost 60% in that same period. Once there were about 30,000 acres of salt marsh in Connecticut, but due to human habitation only about half remains today. While threatening biodiversity, the loss has also disrupted the migration of many birds. Most states have passed laws to protect the marshes from development. Now we must keep man-made pollution from destroying the ones that remain.

7

Shore Birds

A wonderful bird is the pelican,
His bill will hold more than his belican.
He can take in his beak
Food enough for a week,
But I'm damned if I see how the helican.

Dixon Lanier Merritt

Obvious inhabitants of the beach are the ever present birds. Shore birds are amusing as they scamper about, dodging waves or diving into the water to chase fish. They often stand on one leg, the other nestled among the feathers, out of sight. Occasionally they even hop along on one foot. Many shore birds feed in the salt marsh during low tide and return to the sandy beach to search for food at high tide. Many of the birds are only transients, staying for varying periods of time on their way north in the spring or south in the fall. They stop in their migratory flight only for a bite to eat and fly on.

Initially, it may seem very difficult to identify the different species. However, their bills, legs, and behavior can be used to distinguish different types such as plovers, sandpipers, gulls, terns, and skimmers.

The following are but a sampling of the more common shore birds that one might encounter in the habitats covered in this guide.

Plovers have short pigeon-like bills and large eyes. They run with head lowered, then abruptly stop and raise their head. An example that breeds from North Carolina to Massachusetts is the piping plover (*Charadrius melodus*). It is usually whitish in color with a black spot above the forehead, a black ring around the neck, a gray back, a black-tipped yellow bill, and yellow legs and feet. The bird is smaller than a robin (Figure 7-1).

Most sandpipers, unlike plovers, have long, slender bills, often several times longer than the head. The bills can be straight, up-curved or down-curved. Usually sandpipers are more slender and have smaller heads and eyes than plovers. Some species of sandpipers are often seen rushing back and forth in the intertidal zone to find a more promising spot for probing the wet sand with their long bills, always just beyond the reach of the waves.

Size among sandpipers varies considerably. The greater yellowlegs (*Tringa* [*Totanus*] *melanoleucus*) can be more than a foot long. The long, thin bill is dark in color and slightly upturned, and the long legs

Figure 7-1. The piping plover (*Charadrius melodus*) has a distinctive black ring around the neck. (Photo by James P. Mattisson, courtesy of the U.S. Fish and Wildlife Service.)

are bright yellow. Body color is mostly gray above and white below. The lesser yellowlegs (*Tringa* [*Totanus*] *flavipes*) is similar in appearance but the bill is thinner, shorter, and straighter. Also, the bird is slightly smaller, averaging about ten inches in length. Both yellowlegs are more frequently seen on mud flats and in marshes than on sandy beaches. The greater yellowlegs usually wades out into waistdeep water in search of fish and crustaceans, seldom probing into the mud. The lesser yellowlegs preys primarily on insects and crustaceans in shallow pools. Both birds winter in the study area, but the lesser yellowlegs is seldom seen north of Delaware.

The semipalmated sandpiper (*Ereunetes pusillus*) is one of the smallest and most abundant shorebirds along the East Coast. Although it breeds in the Arctic and usually winters south of North Carolina, it migrates through the study area. It is sparrow-size, with a shorter, thicker, bill. Color is brown with dark spots above and white below. The semipalmated sandpiper has black legs, with four toes, which are webbed at the base, on each foot. The least sandpiper (*Erolia minutilla*) is similar in appearance but has yellowish-colored legs. It has four toes, but no webs. The migratory bird passes through the study area, but also winters south of Delaware. The least sandpiper is more frequently encountered in salt marshes and on the mud flats of bays and estuaries.

The most entertaining of the sandpipers is the sanderling (*Calidris alba*). The small animated bird scampers back and forth in the intertidal zone, almost as if playing tag with the advancing and retreating waves. As the bird dashes about on slender legs, it is frantically searching for food with its probing bill. You might imagine that the bird would be better off spending less time dodging waves and more time probing the sand higher in the intertidal zone. However, the game of tag is more efficient because the small organisms it preys upon are closest to the surface just after the water has passed over them as the wave recedes. It feeds on a variety of animals including tiny mollusks and the burrowing mole crab. Occasionally the bird forages for beach hoppers and sand fleas above the intertidal zone. It is difficult to get close enough to see what it is eating. As you approach, it flies down the beach to resume its game of tag with the waves.

The sanderling is smaller than a robin; winter coloration is gray above and white below, with a long, wide, white wing stripe. During the summer, the head, back, and breast are reddish-brown. The coloration blends well with the beach sand. The bill and legs are black in color (Figure 7-2). The bird breeds in the Arctic, but winters in the study area, mostly south of Long Island. It is more frequently encountered from fall to spring, but occasionally during the summer. The small birds, often called peeps because of their calls, were once hunted for food.

Gulls are more abundant by the sea, but are also found around the freshwater Great Lakes and smaller inland bodies of water. One species is found chiefly in the interior and rarely along the sea coast. Thus, the term "sea gull" is inappropriate.

Long, narrow, pointed wings allow gulls to fly and glide gracefully. They have webbed feet and are good swimmers, but seldom dive under the surface. Gulls have slightly hooked bills and are scavengers, usually foraging for food at garbage dumps. Gulls are year-round residents of the study area, and nest in colonies on the ground. Nests are primarily composed of grass.

The adults of different species can be easily distinguished by plumage, but immature gulls are generally brown in color and take about three years to mature.

Figure 7-2. The sanderling (*Calidris alba*) has a black bill and legs. (Photo by Dianne Huppman, courtesy of the U.S. Fish and Wildlife Service.)

The most common gull, the herring gull (*Larus argentatus*), often stands quietly at the water's edge. Average length is two feet, with a wingspan of 4⅔ feet, and a weight of 2½ pounds. It has a yellow bill with a shiny red spot on the lower mandible, and white body color with a gray mantle (Figure 7-3). Adult gulls feed chicks by regurgitating food into their mouths. The red spot on the bill functions as a "releaser" during the initial feedings, because it stimulates the hungry chicks to peck at the bill, stimulating regurgitation. The herring gull is an avid scavenger but also eats crabs and shellfish. The gull will carry a clam high into the air and drop it onto a hard surface, such as pavement, concrete, or an automobile, to break open the shell. Pieces of shell often litter parking areas near a beach. The gull rapidly follows the clam to the ground to keep other gulls from literally stealing the food out from under it. If the clam does not open on the first try, the activity often draws a crowd. It is amusing to watch other gulls flock to the site and

Figure 7-3.
Herring gull
(*Larus argentatus*).
(Courtesy of the
National Park
Service.)

attempt to take the food away from the harassed gull. The herring gull also feeds on the eggs of other shore birds.

Herring gulls breed in the spring, and young chicks take to the air in about six months. They live about 12 years. In the summer, herring gulls are less abundant in the southern region of the study area.

The greater black-backed gull (*Larus marimus*) is similar in appearance to the herring gull, but the mantle is unmistakably black instead of gray (Figure 7-4). The bird winters in the mid-Atlantic states, and breeds from Delaware north.

The ring-billed gull (*Larus delawarensis*) is slightly smaller than the herring gull, and is similar in coloration. However, its yellow bill has a distinctive black ring near the tip. The gull is found more frequently in bays and along sheltered beaches.

The laughing gull (*Larus atricilla*) is much smaller, about half the size of the herring gull. The common name is derived from its laugh-like (ha ha ha ha ha) call. The bill and legs are red, the body is white with a gray mantle, and it has a distinctive dark gray head. In the winter, head feathers are white with a dusky band around the neck. The laughing gull frequently follows and hovers over ferries for free handouts from the passengers. Small pieces of bread tossed into the air will often be caught before striking the water. It feeds primarily on fish, shrimp, and crabs, but also catches dragonflies and other insects in flight (Figure 7-5).

Figure 7-4. The greater black-backed gull (*Larus marimus*) can be distinguished from the herring gull by its darker mantle.

Terns are smaller and more streamlined than gulls, and have deeply forked tails. Gulls fly with their bills pointed straight ahead, while a tern's points down (Figure 7-6). From a hovering position in the air, terns suddenly dive head first into the water to catch fish, usually just beyond the breakers. Because of their flying ability and forked tails, terns are often called sea swallows.

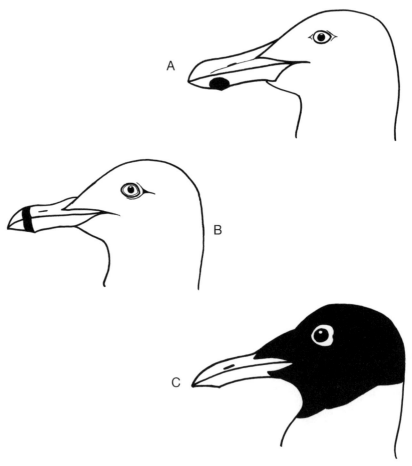

Figure 7-5. The heads of the herring gull (*Larus argentatus*) (A), the ring-billed gull (*Larus delawarensis*) (B), and the laughing gull (*Larus atricilla*) (C).

Figure 7-6. The forked tail of a tern easily distinguishes it from a gull. (Courtesy of the National Park Service.)

The species most frequently seen in the study area is the common tern (*Sterna hirundo*). Body color is similar to that of the laughing gull. The common tern has a white body, a light gray mantle, and a black skull cap in the winter. The feet and bill are bright red, but the bill has a black tip, unlike the laughing gull. The forked tail and black skull cap, instead of the entire head being dark colored, make it easily distinguished from the laughing gull. Common terns aid anglers by flocking above water where large fish are feeding on smaller ones. The terns catch the small fish that are at the surface trying to evade the predator fish. This behavior alerts fishermen to the presence of the large fish.

The little tern (*Sterna albifrons*), also known as the least tern, is smaller in size and has yellow legs and bill. It was almost hunted to extinction in the 19th century; the small birds were slaughtered by the thousands, stuffed, and used to decorate women's hats.

Both terns breed throughout the study area. Terns nest on the ground in colonies but are easily disturbed during nesting, which may inhibit hatching of the eggs. Signs are usually placed by nesting sites to alert beachcombers to keep their distance. If the nesting area is approached, the birds become very agitated. They voice piercing cries of alarm while flying overhead, and often defecate and dive toward intruders.

Skimmers are crow-sized birds that can be easily identified by their behavior and blade-like bills. The lower half of the bill is almost one-fourth longer than the upper. Their legs are so short that when they stand, the body barely clears the sand (Color Plate 15d). Skimmers prefer the calmer water of bays and estuaries where they skim over the surface, with the lower part of the bill cutting through the water to catch surface fish and crustaceans. They are often called cut-waters or shear-waters because of that behavior. When the lower mandible (bill) strikes an object, the upper snaps shut, trapping the prey. Skimmers occasionally feed in the surf zone. They nest on the ground, forming a bowl-shaped hollow in the sand.

The black skimmer (*Rhynchops niger*) breeds in the study area. The body is black above and white below. The legs and bill are red; the bill has a black tip (Figure 7-7).

Figure 7-7. A black skimmer (*Rhynchops niger*) in search of food. When the lower mandible strikes a fish, the upper mandible snaps shut to trap the prey.

Cormorants have webbed feet and are excellent swimmers. While swimming on the surface, their bills are tilted up. The upper mandible is slightly longer than the lower, and the tip of the upper one hooks over the lower one. It is fun to watch them swimming on the surface and see them suddenly dive underwater, where they can cover great distances to catch fish, then wonder where they will surface again. When they surface with a fish, they flip it into the air and swallow the fish head first, avoiding the sharp fin spines. Chinese fishermen train cormorants to dive on command and catch fish for them. The bird is tethered by a line with a noose at the neck preventing it from swallowing the catch. Cormorants often use their wings for propulsion underwater and their tails as rudders. They do not have the oil glands most marine birds have, so their feathers are not waterproof. After leaving the water, they extend their wings to promote drying. Cormorants are found more frequently on the quiet waters of bays and estuaries (Figure 7-8).

When flying, their long necks are extended straight out, but when swimming or standing the neck is "S"-shaped. They often fly in ragged "V" formations, and are mistaken for geese, although they fly much lower than geese. Their nest, usually constructed of sticks and seaweed, is in trees or on rocky ledges.

The double-crested cormorant (*Phalacrocorax auritus*) is found throughout the study area. The body is black with an orange throat pouch.

Figure 7-8. A cormorant has webbed feet and a distinctive hooked bill.

Two interesting shore birds, the brown pelican (*Pelacanus occidentalis*) and the American oystercatcher (*Haematopus palliatus*), are found from Cape Hatteras southward. However, the American oystercatcher is expanding its range and is seen as far north as Cape Cod.

The large familiar pelican can have a wingspan of over six feet (Figure 7-9). The bird has a large, flat bill and an enormous throat pouch that is used underwater like a net to catch fish (Color Plate 16c). Narrow openings in the bill allow the water in the pouch to strain out, leaving fish behind. Although ungainly in appearance on the ground, the pelican is graceful in the air and glides over the water in search of prey. When a fish is sighted, the bird arches up, tucks in its wings, and dives headfirst into the water.

Pelicans have large webbed feet and are good swimmers. However, they do not stay underwater to chase fish, like cormorants. Pelicans have buoyant air sacs under the skin that quickly bring them to the surface and also cushion the body from the impact of hitting the water.

Figure 7-9. The impressive wing span of a brown pelican (*Pelacanus occidentalis*). (Photo by T. C. Maurer, courtesy of the U.S. Fish and Wildlife Service.)

When standing, the long bill rests against the chest. The body is usually brown in color as the common name implies, but in winter the head and neck are white. Pelicans will nest in large colonies in small trees or on the ground, where their nesting is easily disturbed. Nests are composed primarily of sticks and grass. Pelicans are present around Cape Hatteras during the warmer months; they winter farther south.

The American oystercatcher is a year-round resident of Cape Hatteras. It is much smaller than a pelican, growing to about 15 inches. It has a long, flat, pointed, red bill. The body is brown above and white below. The head and neck are black, with an orange ring around each eye; the legs are flesh-colored. (Color Plate 16d).

As its common name implies, the American oystercatcher has a unique way of feeding. When it finds a feeding bivalve with the valves partially open, it sticks the flat bill between the valves and severs the adductor muscles so that the shell gapes open. Then it eats the soft tissues at its leisure. The birds also eat other animals, including shrimp, crabs, and sea urchins. They nest in shallow depressions they create in the sand between the dunes.

The osprey (*Pandion haliaetus*), also known as the fish hawk, visits the study area in the summer. It has the characteristic short, curved bill and sharp talons of a hawk, and is larger than a herring gull, with a wingspan as great as six feet. The wings have a black mark near the tip. The white head has a black streak across each eye and the body is dark above with white below (Figure 7-10).

An osprey will rapidly beat its wings, allowing it to hover high over the water, while it uses its keen eyesight to search for fish. When a prey is spotted, the bird folds in its wings and plunges talons first into the water, sometimes completely submerging, to catch the fish. The name osprey means "bone breaker."

Osprey mate for life and usually nest in trees. People often build wooden platforms and attach them to the tops of long poles, which are set in the ground near the water as osprey nesting sites. The birds construct the nests out of sticks and return to the same nest each year. They can live for more than 20 years. Each year they add sticks to the nest and sometimes the weight crushes the platform (Figure 7-11).

Ducks and geese visit the bays and salt marshes to feed on snails, fish, shellfish, and the green algae *Ulva* and *Enteromorpha*. The famil-

Figure 7-10. The osprey (*Pandion haliaetus*) has the characteristic short, curved bill and sharp talons of a hawk.

Figure 7-11. An osprey (*P. haliaetus*) adds a stick to its nest on top of a man-made wooden platform.

iar mallard (*Anas platyrhynchos*) is a duck frequently seen, especially in the marshes. It has distinctive violet wing-bars. The male has a green head, white neck-ring and brown chest (Figure 7-12). The female is mottled brown in color (Figure 7-13). The wood duck (*Aix sponsa*) also visits the marsh. It often perches in trees and nests in cavities in oak and maple trees, thus the common name. Both sexes have a noticeable crest of feathers on their heads, but the male is more brilliantly colored (Figure 7-14). The male's rainbow iridescence contrasts with the dull colors of the female.

The most common goose in the study area is the Canada goose (*Branta canadensis*). The body is brown, but it is easily recognized by its

Figure 7-12. The male mallard (*Anas platyrhynchos*) has a distinctive green head and white neck-ring.

Figure 7-13. A female mallard with ducklings.

Figure 7-14. The wood duck (*Aix sponsa*) has a distinctive crest of feathers on its head. The male (shown here) has a rainbow iridesence of colors.

black head and neck with a conspicuous white chin-strap (Figure 7-15). The large goose weighs up to 14 pounds and can be 43 inches long. Canada geese fly in large V formations with their necks fully extended.

Shore birds eat insects, snails, worms, crabs, and fish. The long-necked, long-legged, long-billed herons and egrets are frequently seen wading gracefully in the shallow water of marshes and bays in search of fish and other animals. They often stand very still, patiently waiting for a small fish to pass. When a fish is spotted, the long, pointed bill streaks into the water snapping at prey. In flight, their long necks are drawn back in the shape of an S and the neck appears much shorter.

The great blue heron (*Ardea herodias*) breeds throughout the study area; it can be over four feet tall with a wingspan of six feet. The body is bluish-purple and the head partially white with dark plumes extending back (Figure 7-16, Color Plate 16b). It feeds on small fish, eels, snakes, mice, and even small birds.

The little blue heron (*Florida caerulea*) is half the size of the great blue heron. It is darker in color and its head and neck are reddish-purple.

The snowy egret (*Egretta thula*) also breeds throughout the study area. The head has distinctive plumes extending back. It grows to about two feet tall, the body is white, the bill and legs are black, and the feet are yellow (Figure 7-17). The snowy egret feeds on fish, shrimp, crabs,

Figure 7-15. The Canada goose (*Branta canadensis*) is easily recognized by its black head and neck with a conspicuous white chin-strap.

Figure 7-16. The great blue heron (*Ardea herodias*) has a partially white head with dark plumes extending back.

Figure 7-17. The snowy egret (*Egretta thula*) is white with black legs and bill, and its head has distinctive plumes extending back.

frogs, and insects. It uses one foot to stir up the bottom and when a small animal swims away, the bird carefully stalks it.

The common egret (*Casmerodius albus*), which breeds as far north as New Jersey, is twice the size of the snowy egret. It has a yellow bill and black legs and feet, and no head plumes (Color Plate 16a). Both egrets were nearly hunted to extinction for their plumage, which was used for women's hats and hair ornaments. Conservationists stimulated the passage of the Migratory Bird Treaty Act of 1918 that protected the egrets and many other species of migratory birds, saving them from extinction. The common egret feeds on fish, snakes, frogs, and other small animals.

Egrets can be seen from a distance as a spot of white in the green-brown background of the marsh. Herons with darker colored plumage are not as easily seen from a distance. Herons are difficult to approach while feeding, and are easily disturbed. As they fly away they usually voice their irritation. Both egrets and herons nest in colonies, often mixed, on the ground and in large trees. The platform-like nests are built with sticks.

Many of the shore birds are only transients, staying for varying periods of time on their way north in the spring or south in the fall. They stop in their migratory flight only for a bite to eat and fly on. Millions of migratory birds, such as egrets and herons, depend on the bays and marshes for food and rest during their annual migratory cycle.

Bibliography

Auerbach, Paul S. "An introduction to Stinging Marine Life Injuries." *Alert Diver*, January/February 1994.

Ballantine, Todd. *Tideland Treasure*. University of South Carolina Press, 1991.

Barnes, Robert D. *Invertebrate Zoology*. W. B. Saunders Co., 1974.

Berrill, N. J., and Berrill, Jacquelyn. *1001 Questions Answered About the Seashore*. Dover Publications, Inc., 1976.

Black, John A. *Oceans and Coasts: An Introduction to Oceanography*. Wm. C. Brown Publishers, 1986.

Bookspan, Jolie. "Things That Sting." *Alert Diver*, January/February 1994.

Burnett, Joseph W. "Jellyfish Envenomations." *Alert Diver*, January/February 1994.

Carlton, James T. "A Steady Stream of Invading Marine Organisms Creates Ecological Roulette in New England Waters." *Estuarine Research Federation Newsletter*, Volume 19, Number 4, December 1993.

Carson, Rachel. *The Edge of the Sea*. Houghton Mifflin Company, 1955.

Castro, Peter, and Huber, Michael E. *Marine Biology*. Mosby Year Book, 1992.

Coulombe, Deborah A. *The Seaside Naturalist: A Guide at the Seashore*. Simon & Schuster, 1984.

Cruickshank, Allan D. *A Pocket Guide to Birds*. Washington Square Press, Inc., 1960.

Dawson, E. Yale. *How to Know the Seaweeds*. Wm. C. Brown Co., 1956.

Duxbury, Alyn, and Duxbury, Alison. *An Introduction to The World's Oceans*. Addison-Wesley Publishing Co., 1984.

Frings, Hubert, and Frings, Mable. *Concepts of Zoology*. The Macmillan Co., 1970.

Gates, David Alan. *Seasons of the Salt Marsh*. The Chatham Press, 1975.

Gosner, Kenneth L. *Peterson Field Guides: Atlantic Seashore*. Houghton Mifflin Co., 1978.

_____.*Guide to Identification of Marine and Estuarine Invertebrates: Cape Hatteras to the Bay of Fundy*. Wiley-Interscience, 1971.

Gross, M. G. *Oceanography*. Charles E. Merrill Publishing Co., 1980.

Hay, John. *The Sandy Shore*. The Chatham Press, 1968.

Hehre, Jr., Edward J. *A Photographic Guide to the Common Genera of Marine Algae of Eastern Long Island, New York*. Unpublished.

Hickman, Cleveland P., Hickman, Cleveland P., Jr., and Hickman, Frances M. *Biology of Animals*. The C. V. Mosby Co., 1978.

_____.*Integrated Principles of Zoology*. The C. V. Mosby Co., 1974.

Hickman, Cleveland P. *Biology of the Invertebrates*. The C. V. Mosby Co., 1973.

Ingmanson, Dale E. and Wallace, William J. *Oceanography*. Wadsworth Publishing Co., 1985.

Johnson, Ann F. *A Guide to the Plant Communities of the Napeague Dunes, Long Island, New York*. The Mad Printers, 1985.

Kalin, Robert J. *Geology Field Manual*. Moraine Interscience Publishers, 1974.

Knauth, Percy, Editor. *The Illustrated Encyclopedia of the Animal Kingdom*. The Danbury Press, 1971.

Knudsen, Jens W. *Collecting and Preserving Plants and Animals*. Harper & Row, 1972.

Kohn, Bernice. *The Beachcomber's Book*. Puffin Books, 1976.

Leonard, Jonathan N. *Atlantic Beaches: The American Wilderness*. Time-Life Books, 1972.

MacGinitie, G. E., and MacGinitie, Nettie. *Natural History of Marine Animals*. McGraw-Hill Book Co., 1968.

Margulis, Lynn, and Schwartz, Karlene V. *Five Kingdoms: An Illustrated Guide to the Phyla of Life on Earth*. W. H. Freeman and Co., 1982.

McConnaughey, Bayard H., and Zottoli, Robert. *Introduction to Marine Biology*. The C. V. Mosby Co., 1983.

Menninger, Edwin A. *Seaside Plants of the World*. Hearthside Press, 1964.

Meyer, Peter. *Nature Guide to the Carolina Coast*. Avian-Cetacean Press, 1991.

Miner, Roy W. *Field Book of Seashore Life*. G. P. Putnam's Sons, 1950.

Palmatier, E. A. *Key to Common Marine Algae—Field Identification*. Unpublished.

Perry, Bill. *Discovering Fire Island*. Division of Publications, National Park Service, U.S. Department of the Interior, 1978.

Peterson, Roger T. *Peterson Field Guides: Eastern Birds*. Houghton Mifflin Co., 1980.

Petry, Loren C., and Norman, Marcia G. *A Beachcomber's Botany*. The Chatham Press, 1968.

Russell-Hunter, W. D. *A Biology of Higher Invertebrates*. The Macmillan Co., 1969.

_____. *A Biology of Lower Invertebrates*. The Macmillan Co., 1969.

Sea Secrets. The International Oceanographic Foundation, numerous volumes from the mid '70s to the mid '80s.

Seikman, Lula, and Malone, Elsie. *The Great Outdoors Book of Shells*. Great Outdoors Publishing Co., 1965.

Shafer, Thayer, (ed). *New England & the Sea*. Marine Bulletin Number 11, University of Rhode Island, 1972.

Smith, Ralph I., (ed). *Keys to Marine Invertebrates of the Woods Hole Region*. Contribution Number 11, Systematics-Ecology Program, Marine Biological Laboratory, Woods Hole, Massachusetts, 1964.

Soper, Tony. *The Shell Book of Beachcombing*. Taplinger Publishing Co., 1972.

Spitsbergen, Judith M. *Seacoast Life: An Ecological Guide to Natural Seashore Communities in North Carolina*. The University of North Carolina Press, 1983.

Stowe, Keith. *Ocean Science*. John Wiley & Sons, Inc., 1983.

Sumich, James L. *Biology of Marine Life*. Wm. C. Brown Co., 1976.

Taylor, Sally L. and Villalard, Martine. *Seaweeds of the Connecticut Shore*. The Connecticut Arboretum, Bulletin Number 18, 1979.

Teal, John and Mildred. *Life and Death of the Salt Marsh*. An Audubon/Ballantine Book, 1975.

Thorson, Gunnar. *Life in the Sea*. McGraw-Hill Book Co., 1976.

Yonge, C. M. *The Sea Shore*. Atheneum, 1963.

Index

Boldface page numbers indicate illustrations.
Boldface numbers with letters (a, b, c, d) indicate color plates.

251